MW00335573

The Technician's EMI Handbook

The Technician's EMI Handbook

Clues and Solutions

Joseph J. Carr

Newnes

Boston · Oxford · Auckland · Johannesburg · Melbourne · New Delhi

Newnes is an imprint of Butterworth–Heinemann.
Copyright © 2000 by Butterworth–Heinemann

 A member of the Reed Elsevier group

All rights reserved.

No part of this publication may be reproduced, stored in a retrieval system, or transmitted in any form or by any means, electronic, mechanical, photocopying, recording,
or otherwise, without the prior written permission of the publisher.

∞ Recognizing the importance of preserving what has been written,
Butterworth–Heinemann prints its books on acid-free paper whenever possible.

GLOBAL
RELEAF
2000
Butterworth–Heinemann supports the efforts of American Forests and the
Global ReLeaf program in its campaign for the betterment of trees, forests,
and our environment.

Library of Congress Cataloging-in-Publication Data

Carr, Joseph J.
    The technician's EMI handbook : clues and solutions / Joseph J. Carr.
       p.   cm.
    ISBN 0-7506-7233-1 (paperback : alk. paper)
    1. Electromagnetic interference—Handbooks, manuals, etc. 2. Electromagnetic
compatibility—Handbooks, manuals, etc. I. Title.
    TK7867.2. C37   2000
    621.382'24—dc21   00-023808

British Library Cataloguing-in-Publication Data
A catalogue record for this book is available from the British Library.

The publisher offers special discounts on bulk orders of this book.
For information, please contact:
Manager of Special Sales
Butterworth–Heinemann
225 Wildwood Avenue
Woburn, MA 01801-2041
Tel: 781-904-2500
Fax: 781-904-2620

For information on all Newnes Press publications available, contact our World Wide
Web home page at: http://www.newnespress.com

10 9 8 7 6 5 4 3 2 1

Printed in the United States of America

Contents

Preface

Electromagnetic interference (EMI) and electromagnetic compatibility (EMC) are important because of the large number of electronic devices currently in use. Without EMI/EMC considerations, the world would be very different. It would be difficult to operate equipment and devices whenever a transmitter was on the air. Or how about receivers being used in the presence of the noise generated by computers, dimmer switches, and microwave ovens?

We will start this discussion of EMI/EMC by introducing you to EMI/EMC, followed by a discussion of electromagnetic fundamentals and the basics of electromagnetic interference. We will then look at grounding, shielding, and filtering techniques, followed by a discussion of various individual cases. Finally, we will discuss some regulatory issues.

Chapter 1

Introduction to the EMI Problem

All forms of electronic equipment, particularly radio frequency communications equipment, suffer from electrical and electromagnetic interference. To electronic systems such signals constitute a serious form of pollution. The effects may range from merely annoying (e.g., a minor interfering buzz on a radio) to catastrophic (e.g., the crash of an airliner). There are a number of sources (Figure 1.1) of this noise. Some of these can be dealt with at the source; others are beyond our ability to affect at the source and so must be dealt with at the receiver end. The collective term for this pollution is electromagnetic interference (EMI); the ability to withstand such assaults is called electromagnetic compatibility (EMC).

We are all familiar with lightning bolts. The lightning oscillates back and forth between positive and negative ends very rapidly, and is effectively a fast rise-time pulse. As a result, it will have significant harmonics well into the low-band, although the peak is below 500 kHz. Short blasts of static characterize this noise.

The 60-Hz alternating current (AC) power lines are a significant source of noise, especially in the lower frequency bands (including the AM broadcast band). This may seem counterintuitive because those bands are so low in frequency compared with the bands we are discussing. The problem is that the harmonics of 60 Hz extend well into the low-band region of the spectrum. Although they may be down many dozens of decibels from the fundamental, the high voltage and high power levels of the fundamental mean that those "way down" harmonics are still significant to radio receivers at short distances. This situation is seen in the spectrum chart of Figure 1.2.

Several mechanisms are found in 60-Hz AC interference. First, of course, is radiation from the high-voltage distribution lines and the local lower voltage residential feeders. The transformers also radiate signals. If

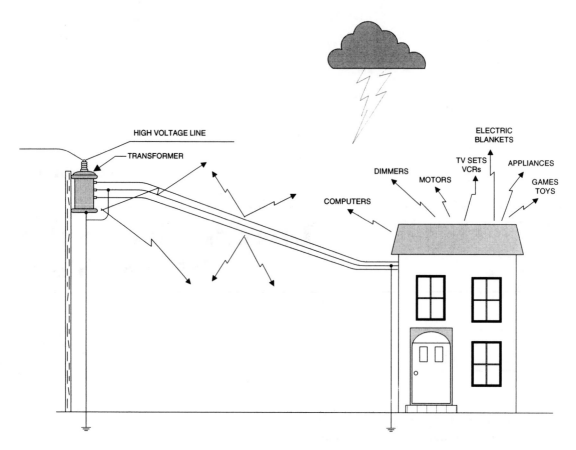

Fig. 1.1 *Typical sources of noise pollution in the environment.*

any of the connections in the electrical circuit are loose or corroded, then the possibility of higher order harmonics increases significantly.

Once you get inside the building, the electromagnetic interference (EMI) situation deteriorates rapidly from room to room. Computers send out very large signals, especially if one is so unwise as to buy unshielded interconnection cables. Light dimmers, microwave ovens, motors on appliances and heating equipment, appliances, and electric blankets all have the potential for creating EMI.

Television sets and VCRs are particularly troublesome in populated areas. In my neighborhood, the housing density is moderately high. I can tell by listening to a high-quality shortwave receiver when a popular television show is being aired. How? Try listening to the low bands! In the United States and Canada, television receivers follow the NTSC color TV standard (which some claim means "Never Twice Same Color"). This means that the horizontal deflection system operates at 15,734 Hz. It is a high-powered pulse with a moderately fast rise time. The harmonics of 15,734 Hz are found up and down the radio dial all the way up to about 20-MHz.

To make the situation worse, the NTSC color subcarrier operates at 3.58 MHz and often has enough power to radiate through poorly shielded television and VCRs. The sit-

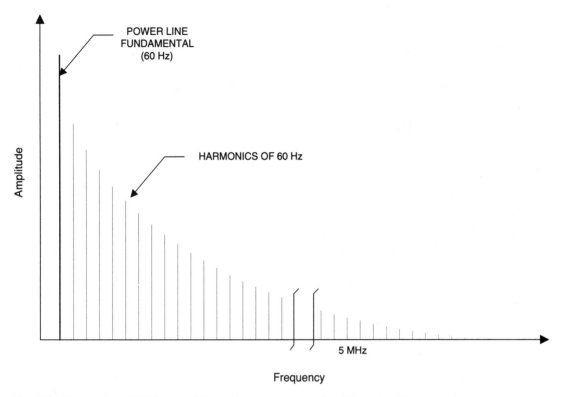

Fig. 1.2 *Harmonics of 60-Hz power line currents can extend well into the RF spectrum.*

uation in countries where the PAL and SE-CAM systems are used may be different, but they are certainly similar.

Just as the sources of EMI/EMC are varied, so are the solutions. In the chapters to follow you will find information concerning the various forms of EMI, the symptoms, and how to overcome the problems caused by EMI. But first, let's take a look at the physical basis for electromagnetic interference.

Chapter 2

Electrical and Electromagnetic Fundamentals

In this chapter the fundamentals of electricity and electrical circuits are presented. For most readers this material is a review of the basics. Those who no longer need instruction on the elementary level are invited to jump ahead to the material that follows. But for others, here is an overview of electrical phenomena that form the basis for EMI/EMC problems.

WHAT IS ELECTRICITY?

We might be tempted at this point to ask, "What is electricity?" No one really knows what electricity is, but we know a great deal about how it behaves. We can observe electricity in action, and we have learned how to control it. Modern technology exists because we know how to control and exploit electricity.

The word "electricity" is derived from the ancient Greek word for the translucent yellow/orange mineral *amber* (fossilized tree resin). The name came to be used for electrical phenomena because the Greeks found that rubbing amber with a cloth produced strange attractive and repulsive forces. Although we now know those forces are from *static electricity*, they truly mystified the Greeks.

Oddly enough, we know little more about the true nature of electricity today than those ancient Greeks who rubbed amber and marveled at the effects. We have, however, made tremendous progress in learning to harness, generate, manipulate, and make practical use of electricity.

Scientists such as Faraday, Ohm, Lenz, and Kirchoff learned much about the effects of electricity and how to use it. Electricity is very predictable in its behavior, and it is that factor that makes it possible to use electricity for high-technology applications. The scientists learned that electricity behaves in a predictable manner, so were able to formulate "laws" that permit the rest of us to predict how electricity will behave.

POSITIVE AND NEGATIVE ELECTRICITY

Many physical phenomena have two opposite attributes, which probably reflects an inherent tendency in Nature toward balance. Magnets have north and south poles, as does the Earth's magnetic field. Electricity also has two natures or "charges": *positive* and *negative*.

In electricity, the positive and negative charges are carried by *atomic particles* (i.e., the parts of the atom). The positive charge is carried by *protons*, while the negative charge is carried by *electrons*. The magnitude of the electric charge carried by protons and electrons is the same, but their polarities are opposite.

There is a large difference in the masses of electrons and protons. The electron mass is about 9.11×10^{-28} grams, while the mass of the proton is 1.67×10^{-24} grams. The proton is about 1,835 times heavier than the electron.

Because protons and electrons carry equal but opposite polarity charges, combining them in a closely related system produces the state of *electroneutrality*, in which the positive and negative charges cancel each other. The positive and negative charges still exist, but to the outside world they appear to be one body with a neutral electrical charge (i.e., no charge).

There is an atomic particle in which the proton and electron are combined. This particle is called the *neutron* and is electrically neutral (as its name implies). The neutron mass is about 1.675×10^{-24} grams.

Every material has both electrons and protons, but because of electroneutrality most do not normally exhibit electrical properties. For example, the skin on your forefinger contains large quantities of both electrons and

protons in a myriad of configurations. Under normal conditions they are in a state of electroneutrality, so do not appear to an outside observer to be electrical in nature. However, if you touch a hot electrical wire, then current will flow and the skin becomes definitely (and painfully) electrical.

Much of what you will learn as you study electronics is what occurs when electrons and protons are not in balance, i.e., when electroneutrality does not exist. Electrical circuits perform work because electroneutrality in that circuit is disturbed. In a dry cell (Figure 2.1), for example, chemical means are used to unbalance the electrical situation inside the cell. As current flows from one terminal to another, through an outside circuit, charge is transferred between materials. When enough charge has passed from one side to the other to establish electroneutrality, the current flow ceases and the battery is considered "dead."

ELECTRONIC MODEL OF THE ATOM

Early Greeks correctly theorized that matter could be broken down into smaller and smaller parts, until at some point the smallest indivisible fraction was reached. A single *atom* is the smallest unit of matter that still retains its properties. The word "atom" comes from a Greek word meaning something that cannot be further divided, or is too small for further division. It is the elementary portion of a material. An atom of hydrogen is still hydrogen and will act like any other atom of hydrogen.

An atom is the smallest unit of a material that cannot be broken into smaller parts by ordinary chemical action. However, the atom is further divisible into subatomic particles: electrons, protons, and neutrons. Chemical reactions are not used for the decomposition of atoms into subatomic particles, but an external force such as electricity or extreme heat can be used to disassociate the electrons and protons of the atoms from each other.

Subatomic particles do not behave like the element they came from. When the hy-

Fig. 2.1   Dry cell.

drogen atom is broken down further, it becomes a single electron and a single proton and does not act like hydrogen any more. These subatomic particles are indistinguishable from electrons and protons that come from other atoms.

Elements

Materials that are formed of a single type of atom are called *elements*. Currently about 106 elements have been identified, although physicists believe that additional elements can be synthesized but are not found in nature. Oxygen, hydrogen, iron, lead, and uranium are examples of elements.

Molecules and Compounds

When two or more atoms are brought together, a *molecule* is formed. Some molecules are of the same material as an atom. For example, oxygen (O) is most often found in diatomic form, i.e., two atoms of oxygen are bound together (designated O_2). The diatomic form of oxygen is an oxygen molecule.

When two or more different types of atoms are bound together, the result is a new material called a compound. Ordinary water is an example of a compound. It consists of two hydrogen (H_2) atoms bound to a single oxygen (O) atom (symbol: H_2O).

Figure 2.2 shows the hierarchy discussed above. Visible matter, such as a drop of water, consists of many molecules or atoms. When it is divided to its smallest fraction that still retains the properties of the visible matter, then it will be either a molecule or an atom depending on the type of matter. When a molecule is further decomposed it forms individual atoms. If the original molecule was an element, then all of the atoms will be the same (e.g., O for diatomic O_2 oxygen), but if it was a compound, then at least two different atoms will be found (two hydrogens and an oxygen in the case of H_2O, water). When the individual atoms are decomposed, they become electrons, protons, and neutrons.

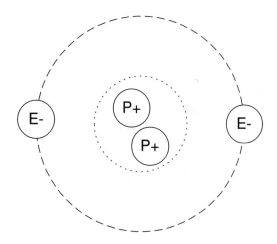

Fig. 2.2 *Simplified helium atom.*

THE BOHR MODEL OF THE ATOM

There are a number of models of how atoms are structured. The one normally used in electronics is the Bohr model, named after Danish physicist Niels Bohr. The Bohr model was formulated in 1911 when Bohr extended work of his mentor, Lord Rutherford. Later work by Max Planck, Albert Einstein, Erwin Schrödinger, George Gamow, Wolfgang Pauli, and Werner Heisenberg, among others, demonstrated that the Bohr model was too simplistic, and their work resulted in the *quantum mechanics* description of the atom. For our current discussion, however, the Bohr model is sufficient. Later, we will extend our discussion to cover electron dynamics and the quantum description. The more complex description is necessary for understanding electron devices such as transistors and integrated circuits, as well as vacuum electron devices used in the microwave region such as magnetrons, traveling wave tubes and klystron tubes.

Bohr's Model

A good analogy for the Bohr model of the atom is our solar system. At the center of the solar system is the Sun. Orbiting the Sun are a number of planets. Figure 2.3 shows various atoms modeled according to Bohr's "solar system" idea. There is a nucleus at the center of

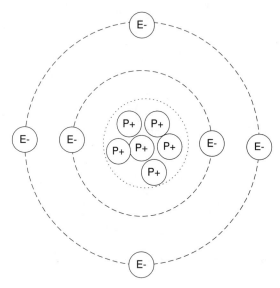

Fig. 2.3 *Atom.*

each atom, consisting of protons and neutrons. Around the nucleus one or more electrons orbit like the planets ("planetary electrons"). Just as gravitational force keeps the planets "captured" in orbit around the Sun, electrical forces keep the electrons bound to the nucleus.

The type of element formed by an atom is determined by the number of protons in the nucleus, and the number of planetary electrons that orbit around the nucleus. For example, the Bohr model for hydrogen (H) consists of one proton in the nucleus and one planetary electron. Helium (He) consists of two protons in the nucleus and two planetary electrons. It also has two neutrons in the nucleus.

Atomic Weight and Atomic Number

The atomic weight of an atom is the total mass of all of the electrons, protons, and neutrons in the atom. The atomic number of the atom is the number of electrons or protons present in the atom. For example, in Figure 2.3 the atomic number is six. Table 2.1 shows the atomic numbers of certain other elements.

Note that each atom has the same number of electrons and protons and so is electrically neutral.

Table 2.1   Atomic Numbers of Selected Elements

Element	Protons	Electrons	Atomic Number
Hydrogen	1	1	1
Helium	2	2	2
Lithium	3	3	3
Oxygen	8	8	8
Fluorine	9	9	9
Neon	10	10	10

ELECTRON SHELLS

The electrons in orbit around the nucleus are distributed into various rings called shells. For each shell there is a maximum number of electrons that can be accommodated. For the first or innermost shell (called the K shell), the maximum number is two. Thus, only hydrogen (H), with one planetary electron, and helium (He), with two planetary electrons, have just one shell. When a shell is filled with its maximum number of electrons, a new shell is created.

Each additional shell has its own maximum number of electrons. The second shell (called the L shell) is completely filled with eight electrons. The additional shells, each further out from the nucleus than the previous shell, are labeled M, N, O, P, and Q. The maximum number of electrons in each shell are as follows:

Shell	Maximum Number of Electrons
K	2
L	8
M	8 or 18*
N	8, 18, or 32*
O	8 or 18*
P	8 or 18*
Q	8

Notes:
*Depending on element

The neon atom has ten electrons in orbit around the nucleus. Two fill the innermost shell, while the remaining eight completely fill the next higher shell. The configuration of eight electrons in a shell forms what is called a *stable octet*. Such atoms do not easily combine with other atoms and so are considered chemically inert. The inert gases are helium, neon, argon, krypton, xenon, and radon. All of these elements have a completely filled outer shell and so do not easily react with other elements. The elements other than neon, however, have outer shells that are not filled and so are chemically active.

The total number of electrons that can be accommodated in the first four shells can be found from

$$n = 2N^2 \qquad (2.1)$$

where:

> n is the maximum number of electrons in the shell
> N is the shell number (1, 2, 3, 4)

ELECTRON SUBSHELLS AND ELECTRON SPIN

Each shell in the atom, except the K shell, is divided into two or more subshells, labeled *s*, *p*, or *d*. The subshells are designated by the subshell letter appended as a subscript to the shell letter (e.g., M_d designates the *d* subshell of the *M* shell).

The distance of the shell from the nucleus defines the energy level of the electrons. It is a fundamental requirement that no two electrons can have exactly the same description. If they are in the same orbit, then they must have different spin directions (for our purposes, the spins are clockwise and counterclockwise). Consider the iron (Fe) atom, for example (Table 2.2). Iron has 26 electrons orbiting around a nucleus of 26 protons. Each shell except K and N has two or more subshells. The K shell does not have subshells (even though the "K_s" designator is often used). The N shell has only one subshell (N_s) because there are only two electrons in this shell, and they are distinguished by having opposite spins.

Table 2.2   Electron Shell Structure for Iron

Shell	Subshell	CW* Electrons	CCW** Electrons
K		1	1
L	s	1	1
L	p	3	3
M	s	1	1
M	p	3	3
M	d	5	1
N	s	1	1

Notes:
*CW = clockwise
**CCW = counterclockwise

ELECTRON VALENCE AND FREE ELECTRONS

The apparent properties of the atom seen by the outside world are determined by the electrons in the outer shell. If the outer shell is completely filled, for example, there is little possibility of chemical reaction with other atoms. The valence of the atom determines how easily it will attract or give up electrons from the outer shell. The outer shell is therefore called the *valence shell*, and the electrons in the outer shell are called *valence electrons*.

A completely filled outer shell has a valence of zero. But if the outer shell is not filled, it will have a valence determined by the number of electrons present. The goal of the atom is to form a stable octet in the outer shell. Consider the copper atom. It has an atomic number of 29, so the innermost shell has 2 electrons, the next shell has 8 electrons, and the third shell has 18 electrons. All of these shells are completely filled, but there is one electron left over, so it goes into the outer shell. That electron is the valence electron for copper.

There are two ways of specifying valence. Because copper (Cu) has one electron in the outer shell, it is said to have a valence of +1. However, it is also possible to describe valence in terms of the number of electrons needed to fill a stable octet. In the case of copper, that number is 7, so the valence of copper can be described as either +1 or –7.

Elements with similar valence numbers behave in a similar manner in electrical circuits and chemical reactions. If an atom has a high number of electrons in the outer shell, such as 6 or 7, the electrons are bound relatively tightly to the nucleus, although not as tightly as in a stable octet. It takes a large amount of external energy to break an electron loose from the outer shell of such an atom.

In atoms with few electrons in the outer orbit (e.g., Cu with 1), the electrons are less tightly bound and can be broken free relatively easily. Electrons that are disassociated from their atoms are called mobile electrons, or more commonly free electrons. It is the free electrons that are responsible for electrical current.

Ions

Electrically charged particles are called ions. An electron, for example, is a negative ion, while a proton is a positive ion. However, there are other ways to produce the electrical charge needed for an atom to be called an ion. For example, if an atom takes on an extra electron, it is no longer electrically neutral because the number of electrons and protons is not balanced. Such an atom has an electrical charge of –1 unit and is a negative ion. Similarly, if an atom loses an electron, there are more protons than electrons, so the net electrical charge is +1; such an atom is a positive ion. Atoms with one too many or too few electrons behave electrically like point charges, even though they have a mass of approximately the entire atom.

CONDUCTORS, INSULATORS, AND SEMICONDUCTORS

Three types of materials are commonly used in electronics: conductors, insulators, and semiconductors. The distinction between these categories is found in the valence band.

If an element has few electrons in the valence band, then they are easily disassoci-

Table 2.3   Conductors

Element	Symbol	Atomic Number	Valence
Silver	Ag	47	+1
Copper	Cu	29	+1
Gold	Au	79	+1
Aluminum	Al	13	+3
Iron	Fe	26	+2

ated from the atom to become free electrons. Such elements are conductors. Copper and the other metals are examples of conductors because they have fewer electrons in the valence shell.

Because of the large number of free electrons, conductors easily pass electrical currents. Copper, for example, is a very good conductor even at room temperature. Ordinary thermal agitation energy is enough to dislodge a large number of free electrons in copper.

Metallic wires and printed circuit tracks are used to carry electrical currents in circuits. Silver (Ag) is the best conductor, but is terribly expensive for use as wires. Table 2.3 shows several conductors in descending order of conductivity.

Although copper is less of a conductor than silver, it is preferred in electronics because of cost. Gold is sometimes used when it is necessary to guard against tarnishing, such as on electrical connectors and some printed circuit boards. Aluminum is also used extensively in electrical power distribution and on the internal connections of certain electronic components. Because it cannot be soldered, aluminum must be used in applications where mechanical means are used to make electrical connections. Aluminum is lighter and usually cheaper than copper, so it is a good candidate for cross-country power transmission lines.

Insulators have a larger number of electrons (closer to eight) in the outer shell, and so do not easily give up electrons to become free electrons. Because they have few free electrons, the insulators do not support the flow of electrical current. Because they do not support current flow, insulators are used to isolate or prevent the flow of electricity where it is not desired. Wires are often covered with an insulating material so that electrical current carried in the wire does not go elsewhere ("short circuit"). Examples of materials that make good insulators are ceramic, glass, dry wood, plastic, and rubber.

Semiconductors are in the middle of the valence scale between insulators and conductors. These elements (carbon, silicon, germanium) have four valence electrons each and so will support the flow of current with moderate amounts of external energy applied, but not so much as conductors. Carbon, for example, can be used as an electrical conductor, but is not nearly as good a conductor as copper or the other metals.

THE UNIT OF ELECTRIC CHARGE

Recall that amber produces some "strange" repulsive or attractive forces when rubbed. Other materials also produce this same effect: hard rubber, animal fur or hair, and others. If small bits of paper are brought near the rubbed object, they are attracted to it. If a hard rubber rod is rubbed against animal fur, the mechanical work of rubbing the two materials together is sufficient to cause a transfer of electrons from the fur to the rubber rod.

Electricity observed in this manner is called static electricity because it is not in motion. Ordinary experiences can create very large amounts of static electricity. If you walk across a woolen rug, or wear a woolen sweater, then you will sometimes experience a sharp little electrical shock when you touch another object. The field of study of static electricity is called electrostatics.

In 1785, the French physicist Coulomb gave us a means for expressing the amount of electrical charge. The basic building block is the atomic unit of charge, which is the amount of electrical charge present on a single electron or single proton. Because electrons and protons have equal charges, but

opposite polarities, the atomic units of charge are designated −1 for electrons and +1 for protons.

The basic unit for electrical charge is the coulomb (C), which is defined as 6.242×10^{18} atomic units, or put another way, the sum of the electrical charges of 6.242×10^{18} electrons or protons. Sometimes electrical charge is designated by the letter "Q," with −Q being used for a charge produced by electrons and +Q for charges produced by protons.

We can also now express the basic atomic unit of charge in terms of coulombs by taking the reciprocal of 1 coulomb. Thus, the charge on any one electron or proton is $1/(6.242 \times 10^{18}) = 1.602 \times 10^{-19}$ C.

ELECTRICAL POLARITIES

There are two electrical polarities: negative and positive. Because of historical convention, we arbitrarily assign the negative (−) sign to the type of electrical charge found on electrons, and the positive (+) sign to the charge on protons.

All charged particles generate electrical lines of force around themselves. These lines represent the electric field of the charge. We find two rules in effect whenever two charges are brought into close proximity with each other:

1. Like charges repel each other

2. Unlike charges attract each other

In Figure 2.4A two positive charges are in close proximity to each other, so their electrical fields repel each other, and the charges are forced apart. The same result also occurs with two negative charges (Figure 2.4B). In Figure 2.4C, on the other hand, we see unlike charges in close proximity, one positive and one negative. The electrical force in this situation is attractive, so the particles move closer together. Because the electron is so much lighter than the proton, to the outside observer it appears as if the proton were stationary and the electron moves to or away from the proton as dictated by the polarities.

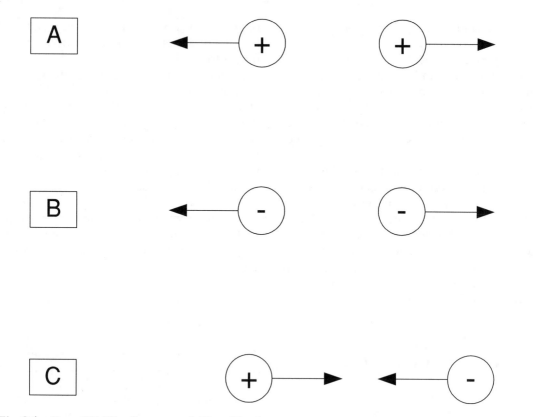

Fig. 2.4   *(A and B) Like charges repel; (C) unlike charges attract.*

The force is represented by the electrical field lines in Figure 2.5. Coulomb's law describes the force as having a value of:

$$F = \frac{1}{4\pi\varepsilon_o} \times \frac{Q_1\,Q_2}{r^2} \qquad (2.2)$$

where:

F is the attractive or repulsive force in newtons

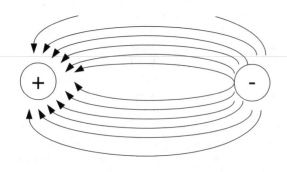

Fig. 2.5   *Electric force field between charges.*

ε_o is the permittivity of free space $(8.8543 \times 10^{-12}/\text{N–m}^2)$

Q_1 and Q_2 are the electrical charges in coulombs

r is the distance between the charges

Because the quantity $1/4\pi\varepsilon_o$ is a constant, equal to about 9×10^9 N-m^2/C^2, we can simplify Equation 2.2 in the form:

$$F = \frac{9 \times 10^9 \, Q_1 \, Q_2}{r^2} \qquad (2.3)$$

Note that the distance between charges, r, is in the denominator and is squared. Because of this relationship, we know that electrical charges obey the inverse square law. Doubling the distance between charges reduces the force to one-fourth the original force. Similarly, halving the distance results in a force that is four times as large.

MILLIKAN'S OIL DROP EXPERIMENT

The basic electronic charge was calculated by American physicist Robert A. Millikan (1868–1953) in his famous "oil drop experiment" conducted between 1909 and 1917. Millikan won the Nobel Prize in physics for this experiment. He created an evacuated chamber into which oil droplets could be sprayed from an atomizer. The friction of the atomizer caused some electrons to be stripped free, while an X-ray source ensured that other electrons would be freed.

Millikan arranged a pair of electrodes and a voltage source such that the oil droplets were inside an electrical field. Normally, gravity would force the oil droplets to fall toward the Earth. Counterbalancing the gravitational attraction, however, was the electrical force. By placing the positive side of the electrical source on the upper plate, Millikan could attract the electron charged oil droplets upward. He observed the oil droplets through a microscope, while adjusting the voltage between the plates until the downward force of gravity was exactly matched by the upward force of the electrical field, suspending the droplet, motionless, in the space between the plates. Since both the force of gravity and the electric field strength are known, the electronic charge could be calculated. He noted that the values tended to be integer multiples of 0.16×10^{-18} C (i.e., 0.32, 0.48, etc), and so concluded that the elementary charge is 0.16×10^{-18}, or as it is usually written 1.6×10^{-19} C (later refined to 1.602×10^{-19} C).

ELECTRICAL POTENTIAL

Electrical potential refers to the ability to do work, specifically the work of moving electrical charges from one place to another. Potential is created by a difference in the amount of electrical charge at the two points. In Figure 2.6A two regions of electrical charge are shown, each designated with +Q symbols. Recall that a +Q charge will attract electrons, which are negatively charged. Also recall that a –Q charge will repel electrons. For the sake of this discussion the drawing is simplified in that we will assume that one unit of negative electrical charge is attracted or repelled by one unit of Q.

In the situation shown in Figure 2.6A, region "A" has a charge of +Q = 4, while region "B" has a charge of +Q = 2. Because of our rule that each unit of +Q attract one unit of negative charge, we see that four negative charge units are attracted to "A" and two units are attracted to "B." The difference is a net flow of two units to "A," i.e., (4) – (2) = 2. To an outside observer, there is a flow of negative charge of two units toward region "A."

A slightly different situation is shown in Figure 2.6B. In this scenario, region "A" is charged +3Q, and region "B" is charged –1Q. The potential difference is (+3Q) – (–1Q) = +4Q. Three electrons are attracted to region "A" by its charge of +3Q, and another unit is repelled by region "B," resulting in a net exchange of 4 units toward "A."

Note that the same situation is found in both scenarios: A difference in electrical potential exists between "A" and "B," so there is an unbalanced condition. The flow of electrons attempts to restore the neutral balance. The flow of electrons forms an electrical current, which can only exist between two regions of an electrical potential difference. Another term for potential difference is electromotive force (emf).

The unit of measure for potential difference is the *volt*, named after Alessandro Volta (1754–1827), who did much pioneering research work in electricity and invented a rudimentary chemical emf source called a voltaic cell. When you examine a carbon–zinc dry cell and find that it is rated at 1.5 volts, this rating tells you that there is an electrical potential difference of 1.5 volts between the positive and negative terminals when the battery is fully charged.

The volt is symbolized by the letter "V" when the value is being indicated, although in some formulas "*E*" is used to represent voltage. It is sometimes confusing when V and *E* are used interchangeably, even though

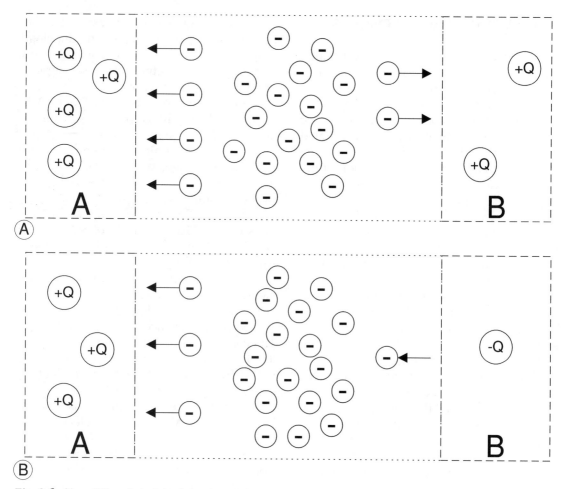

Fig. 2.6 *Two different electric charge scenarios.*

they mean the same thing. In this text V is used to indicate units, while *E* is used in equations, i.e., in the form *E* = 1.5 V.

The volt is defined as the potential required to move one coulomb of charge (i.e., 6.25×10^{18} electrons) with one joule (1 J = 0.7375 ft-lbs) of work. In other words,

$$1 \text{ volt} = \frac{1 \text{ joule}}{\text{coulomb}} \qquad (2.4)$$

There are a number of different ways in which the charge at two different points can become unbalanced. In a battery or voltaic cell, for example, the difference in potential is created chemically. In generators and alternators, mechanical and magnetic means are used. In the atmosphere, lightning can exist because a huge potential difference arises because of friction creating different numbers of charged ions. When the potential difference, i.e., the difference in the number of ions, exceeds a certain critical point, current flows between the two regions and is seen from the ground as lightning.

The volt is commonly used in electrical power and electronics circuits. In some cases, however, the volt is either too large or too small for practical use, so metric prefixes are commonly employed to modify the unit. For example, a kilovolt (kV) is 1,000 volts, while a millivolt (mV) is 1/1,000 of a volt (i.e., 1 mV = 0.001 V). The standard voltage units are shown in Table 2.4.

Table 2.4   Standard Voltage Units

Unit	Symbol	Value
Nanovolts	nV	0.000000001 V
Microvolts	μV	0.000001 V
Millivolts	mV	0.001 V
Kilovolts	kV	1,000 V
Megavolts	MV	1,000,000 V

ELECTRICAL CURRENT AND ITS UNITS

Electrical current is the flow of electrical charge between two points. If there is no potential difference, the free electrons in a conductor will exhibit random motion. Although individual electrons may flow in a particular direction for a short distance (which is the nature of "random" motion), they are cancelled out by other electrons moving in the opposite direction. The net transfer of charge throughout the conductor is zero, so no current is seen to exist. Current can only exist when there is a potential difference, or voltage, between the two points. The magnitude of the flow is proportional to the magnitude of potential difference.

The basic unit of current is the ampere (A), which is named for French physicist André M. Ampère (1775–1836). It is common practice to abbreviate ampere as amp or amps. The ampere is defined as the flow of 1 coulomb (6.25×10^{18} electrons) of charge past a given point in 1 second, or

$$1A = \frac{1 \, \text{coulomb}}{\text{second}} = \frac{6.25 \times 10^{18} \, \text{electrons}}{\text{second}} \quad (2.5)$$

An electrical conductor containing a large number of free electrons has an electrical potential difference (V) between the $-Q$ and $+Q$ ends. A metering device is in the line to count the electrons as they pass point P. One ampere (1 A) of current exists when the meter reads 6.25×10^{18} electrons flowing past point P per second. If the number of electrons doubles to 1.25×10^{19} electrons per second, then a current of 2 A exists.

If the potential difference, V, drops to zero volts, then there can be no current flow. Electrons resume their random motion and there is no net flow in any single direction.

The symbol for current is either I or i. Later you will learn that in some situations there is a difference in the usage of these symbols, but for the time being we will use only the uppercase I.

Instruments that measure current are called ammeters, milliammeters, or microammeters, depending on the maximum level of current that the device will measure.

ELECTRICAL CURRENT VS ELECTRICAL CHARGE

Some people confuse the terms electrical current and electrical charge. The concept of charge refers to a quantity of electricity and is normally meant to refer to the static electricity in a dielectric (i.e., an insulator material). Charge is measured in coulombs, which is defined as a specific number of charge carriers (e.g., electrons). Current, on the other hand, is charge in motion. A current exists in metals and other conductors, semiconductors, and certain liquids and gases. We can define current and charge in terms of each other using the expression:

$$Q = I \times T \quad (2.6)$$

where:

Q is the electric charge in coulombs (C)
I is the current in amperes (A)
T is the time in seconds (s)

From this expression we can derive any of the three variables by suitable rearrangement:

$$I = Q/T \quad (2.7)$$

and

$$T = Q/I \quad (2.8)$$

TYPES OF CURRENT FLOW

Thus far we have discussed the electron flow model of current. There are actually four current types: electron, ionic, hole, and successive depolarization.

Electron Flow

This type of current is the type that has been discussed in this chapter up until now. Until recently, many electronics professionals could spend their entire careers only needing to know about this type of current. The basic unit of electron flow is the elementary electric charge, Q_e, which has been defined earlier as 1.602×10^{-19} coulombs. Q_e has a negative polarity. Electron flow is found in metallic and other conductors, in vacuum tube devices, and in N-type semiconductors.

Semiconductors each have four valence electrons and are neither good insulators nor good conductors. By adding a dopant with either a surplus of free electrons or a deficiency of free electrons, the semiconductor material can be converted to either N-type (surplus of electrons) or P-type (deficiency of electrons).

Ionic Flow

Ions can be individual electrons, or they can be atoms or molecules that are not electrically neutral so carry a net electric charge. In the case of electrons, the current flow is as described in the Electron Flow paragraphs above. If an atom or molecule has at least one additional electron, then it has a net negative electrical charge. Similarly, if it is deficient by one electron, then it carries a net positive charge. In either case, the magnitude of the electric charge is $\pm N Q_e$, where N is an integer number (e.g., 1, 2, 3, . . .). The sign is positive or negative depending on whether the net charge is positive or negative. Ionic flow is common in liquids, gases, and living tissues. Because ions have the massive nucleus at the center of the atom, they are a lot less mobile than free electrons.

Hole Flow

The concept of *holes* is used extensively in semiconductor theory. Although it causes

some confusion among students at first, it is really quite simple: a hole is a place where an electron should be, but isn't. If the valence ring of an atom is missing a required electron, then there is a net positive charge on that atom. The spot where the missing electron should be is the hole. The hole can be treated mathematically as if it were a positive charge with the same mass as an electron. If an electron is captured from a nearby atom, that "hole" is filled, but at the expense of creating a new hole in the donor atom. Although it was an electron that actually moved, to the outside observer it would appear as if the hole moved. Hole flow is found in P-type semiconductors.

Successive Depolarization

This form of current flow is found in the nervous systems of humans and other mammals. Biological cells are essentially miniature batteries with a –90 mV potential difference, with the inside more negative than the outside. Certain cells, such as those of the nervous system, when stimulated will depolarize. In this condition they have a +20 mV potential. In a short period of time (a few milliseconds) the cell repolarizes to –90 mV. When it depolarizes, however, the reversal of polarity triggers the next cell in line to depolarize, and then the next . . . and so forth until the current reaches the end of the nerve. One of the dangers of electricity is that this natural electrical system is overcome by the external electrical energy, causing injury or death to the victim.

RESISTANCE TO ELECTRICAL CURRENT

If you pass an electrical current through a conductor and measure the conductor's temperature, you will notice a rise in temperature as the current flows. This effect is so pronounced that a method used in standards laboratories to calibrated current measuring devices is to measure the heat dissipated as the current flows. From the Law of Conservation of Energy we can deduce that the heat dissipated in an electrical conductor indicates that there is some form of opposition to the flow of current. That opposition is called electrical resistance, or simply resistance. Resistance is symbolized by the letter "R."

The unit of resistance is the ohm, named after German physicist Georg Simon Ohm (1747–1854), and symbolized by the Greek uppercase letter "omega" (Ω). The ohm is defined in terms of the resistance that exists when a current of 1 ampere, over a duration of 1 second, produces a heat dissipation of 0.24 calories:

$$1\Omega = \frac{1A}{1s} \text{ for } 0.24 \text{ calories of heat} \quad (2.9)$$

If there were no resistance in a circuit, then any electrical potential at all would produce destructively large electric currents. The resistance serves to limit the current that can flow under the influence any given value of potential.

The ohm can also be defined as the value of resistance that exists when 1 volt of potential produces 1 ampere of current (i.e., 1 Ω = 1 V/1 A).

CONDUCTANCE

Whereas resistance (R) describes the opposition to the flow of current, the conductance (G) describes the ability of a material to conduct electricity. Conductance is the reciprocal of resistance, so we may write:

$$G = 1/R \quad (2.10)$$

and

$$R = 1/G \quad (2.11)$$

At one time the unit of conductance was called the mho, which is "ohm" spelled backwards. The symbol for the mho was the Ω turned upside down. Today, however, the unit of conductance has been renamed the siemens (S), after the brothers William and Ernst von Siemens, early electrical pioneers and inventors. In practical terms there is no difference because 1 siemens equals 1 mho.

CURRENT FLOW DIRECTION

There is some confusion over the direction of the flow of current: some say it flows from positive to negative (called "conventional flow"), while others say it flows from negative to positive ("electron flow").

Part of the problem is that the positive-to-negative conventional flow was a guess made by early researchers (it is attributed to Benjamin Franklin), and they got it wrong. Another reason for the difference is subtle differences in how "positive" and "negative" are defined. In either case, the charge carriers are electrons and current flows from a region of excess electrons to a region of electron deficiency.

In conventional flow, the region of excess electrons was called positive because it has more electrons than the region with an electron deficiency. Similarly, the deficiency region is labeled negative because it has too few electrons compared to the other region. Because the flow is from excess charge toward deficient charge, the conventional flow definitions compel a flow from positive to negative.

In electron flow, the labels positive and negative are given to the sign of the electrical charge present in the two regions. If a region has more electrons than another region, then it is negative with respect to the other region because of the difference in charges.

For example, suppose one region has a charge of –2Q and another region a charge +1Q, resulting in a potential difference of –1Q. This means that there are more electrons, carrying negative charges, in the –2Q region and fewer in the +1Q region. Charge, i.e., electrons, will travel from the –2Q region toward the +1Q region until the net difference of charge between them is zero. When that occurs both regions carry the same electrical charge, so there is no potential difference and no current flow.

Figures 2.7A and 2.7B show both conventional and electronic current flow. In both cases a load (R) is connected across a voltage source (V), giving rise to a current (I). Conventional flow is shown in Figure

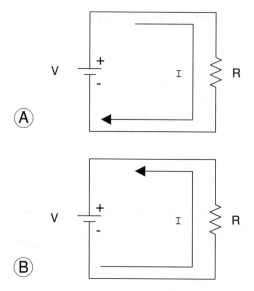

Fig. 2.7 *(A) Simple series circuit showing conventional current flow, and (B) showing electron flow.*

2.7A and proceeds from positive to negative; electronic flow is shown in Figure 2.7B and proceeds from negative to positive.

In straight electrical circuits the difference between electron and conventional flow is not important because all of the equations still work. In electronic devices, however, either direction can still be used but the descriptions of how devices work are easier to understand when the flow is described as negative-to-positive electron flow.

ELECTRICAL POWER

Electrical power is an indication of the work done by electricity. The unit of power is the watt, which is named for Scottish inventor James Watt (1736–1819). In electronics additional units are also often used: kilowatt, milliwatts, microwatts, and so forth.

Kilowatt	1000 watts
Milliwatt	1/1000 watts
Microwatt	1/1,000,000 watts

The watt is defined as the work required to move 1 coulomb of charge in 1 second under the influence of a potential difference of 1 volt. You can calculate the number of watts dissipated in an electrical circuit by finding the product of the current and voltage:

$$P = V \times I \qquad (2.12)$$

where:

P is the power in watts (W)
V is the potential difference in volts (V)
I is the current in amperes (A)

ELECTRICAL SOURCES

Voltage and current can be generated by any of a number of different methods. Which is used in any given case is determined by the nature of the use being made of the electricity.

Static Electricity (Friction Generated)

Static electricity is generated by friction between two insulating materials. When the materials are rubbed together, electrons are removed from atoms within the materials, giving rise to a static electric charge. In practical electronics this source of electricity causes tremendous problems. A person walking across the floor can build up a charge of several thousand volts. Some circuits will be damaged with these potentials.

Mechanical Deformation

Certain crystalline materials (e.g., natural quartz) have a property called piezoelectricity, i.e., the generation of electrical potentials when the crystal is mechanically deformed.

Conversion of Chemical Energy

Cells and batteries are used to produce current and voltage by immersing two electrodes of different material in a chemically active wet or dry electrolyte. Both electrodes react chemically with the electrolyte, but one will give up free electrons while the other accepts free electrons. As a result, one electrode will accumulate a positive charge, while the other accumulates a negative charge. This action creates a potential difference between the electrodes. When an external circuit is connected, allowing the transfer of charge, the electrodes will eventually become neutralized. Under this condition the cell or battery is no longer able to produce current or a potential difference (it is dead).

Magnetic and Electromagnetism

Magnetism and electricity are closely interrelated with each other. When an electrical charge moves, i.e., when current flows in a conductor (Figure 2.8), a magnetic field is created surrounding the conductor. Similarly,

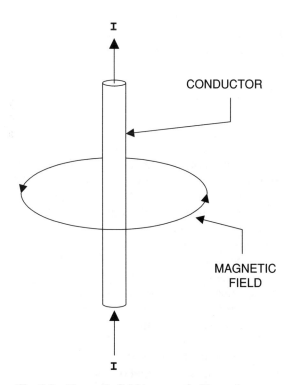

Fig. 2.8 *Magnetic field is generated by a charge in motion.*

if a magnet moves such that its magnetic field intercepts the conductor, a current will be created in the conductor. It doesn't matter whether the conductor moves or the magnet moves (or they both move), as long as they move relative to each other. Under this circumstance current is produced in the conductor by the motion of the magnetic field. This phenomenon is the basis for power generation, but it's also the basis for many forms of EMI problems!

THE ELECTROMAGNETIC FIELD

Radio signals are electromagnetic (EM) waves exactly like light, infrared, and ultraviolet, except for frequency. The EM wave consists of two mutually perpendicular oscillating fields traveling together. One of the fields is an electric field while the other is a magnetic field.

In dealing with both antenna theory and radio wave propagation, we sometimes make use of a theoretical construct called an isotropic source for the sake of comparison and simpler arithmetic. This same concept is useful in EMI/EMC. An isotropic source assumes that the radiator (i.e., "antenna") is a very tiny spherical source that radiates equally well in all directions. The radiation pattern is thus a sphere with the isotropic antenna at the center. Because a spherical source is uniform in all directions, and its geometry is easily determined mathematically, signal intensities at all points can be calculated from basic principles.

The radiated sphere gets ever larger as the wave propagates away from the isotropic source. If, at a great distance from the center, we take a look at a small slice of the advancing wave front, we can assume that it is essentially a flat plane, as in Figure 2.9. This situation is analogous to the apparent flatness of the prairie, even though the surface of the Earth is a near-sphere. We would be able to "see" the electric and magnetic field vectors at right angles to each other (Figure 2.9) in the flat plane wave front.

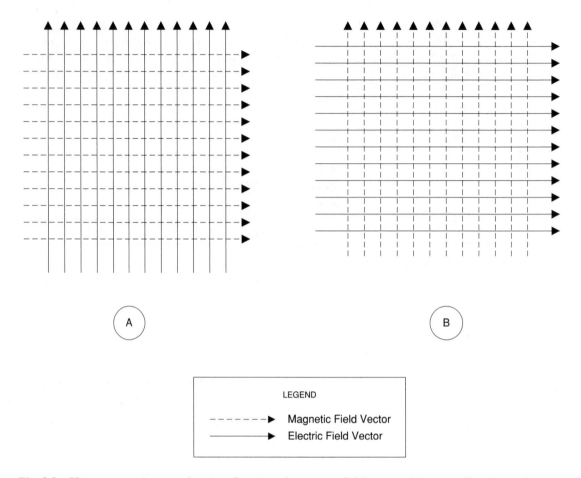

Fig. 2.9   *Electromagnetic wave showing electric and magnetic field vectors: (A) vertically polarized; (B) horizontally polarized.*

The polarization of an EM wave is, by definition, the direction of the electric field. In Figure 2.9A we see vertical polarization because the electric field is vertical with respect to the Earth. If the fields were swapped (Figure 2.9B), then the EM wave would be horizontally polarized.

These designations are especially convenient because they also tell us the type of antenna used: vertical antennas (as are common in landmobile communications) produce vertically polarized signals, while horizontal antennas produce horizontally polarized signals. Some texts erroneously state that antennas will not pick up signals of the opposite polarity. Such is not the

case, especially in the HF and lower VHF regions. A loss of approximately 20 to 30 dB is observed due to cross-polarization.

An EM wave travels at the speed of light, designated by the letter c, which is about 300,000,000 meters per second (or 186,000 miles per second if you prefer English units). To put this velocity in perspective, a radio signal originating on the Sun's surface would reach Earth in about 8 minutes. A terrestrial radio signal can travel around the Earth seven times in 1 second.

The velocity of the wave slows in dense media, but in air the speed is so close to the "free space" value of c that the same figures are used for both air and outer space in practical problems. In pure water, which is much denser than air, the speed of radio signals is about 1/9 the free space speed. This same phenomena shows up in practical work in the form of the velocity factor (V) of transmission lines. In foam dielectric coaxial cable, for example, the value of V is 0.80, which means that the signal propagates along the line at a speed of 0.80c, or 80 percent of the speed of light.

SOURCES OF ELECTROMAGNETIC INTERFERENCE

The sources of electromagnetic interference (EMI) are the electrical, magnetic, and electromagnetic fields that emanate from the wiring of transmitters, AC power mains, and anything that runs on electricity. These fields are filterable, containable, and shieldable . . . that's what this book is all about.

Chapter 3

Fundamentals of Electromagnetic Interference

Before beginning our study of electromagnetic interference (EMI) and electromagnetic compatibility (EMC) we need to lay down a few fundamentals. We need to know these things in order to diagnose and eliminate the problems in equipment.

FUNDAMENTAL CAUSES OF EMI

The fundamental causes of EMI problems can be broken into three categories:

1. Fundamental overload

2. Intermodulation

3. Spurious emissions from a transmitter

Fundamental Overload

Fundamental overload is overload of the affected equipment by the transmitter's fundamental frequency. The overload may be on-channel or off-channel (or channels may not be involved at all!). The suscepti-

ble equipment must be able to do two things:

1. Respond properly to desired signals

2. Not respond to undesired signals (Figure 3.1)

Fundamental overload involves equipment that is not able to meet the second of these criteria. The receiver or other equipment responds to undesired signals, caus-

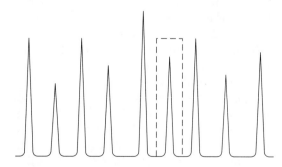

Fig. 3.1 *An electronic device should do two things: respond only to desired signals, and not respond to undesired signals.*

ing problems. The emphasis here is on the susceptible equipment, as the transmitter is legally operated.

Intermodulation

Whenever two frequencies are put together in a nonlinear manner, we get product frequencies that are equal to $mF_1 \pm nF_2$, where F_1 and F_2 are the frequencies, and m and n are integers or zero. The nonlinearity can be any PN diode junction, such as exist in transistors and ICs. Or they can be natural and man-made "diodes" formed by having two dissimilar metals or semiconductors in contact with each other ("rusty bolt effect").

Transmitter Spurs

A transmitter produces a fundamental signal that is on the operating frequency. Unfortunately, they also produce spurious frequencies (called "spurs") that they are not licensed to operate on. These spurs can be harmonics (exact integer multiples of the fundamental frequency), noise, mixer products, parasitic oscillations, or the harmonics of local oscillator frequencies.

The noise is wideband noise, phase noise, and other noise products. Mixer products are due to the effects of mixers, and follow the $mF_1 \pm nF_2$ rule. Parasitic oscillations do not usually have any relationship to the transmitted frequency, as they are due to such things as the resonant frequency of an RF choke, or something similar. The local oscillator products are found in those transmitters (mostly VHF and UHF) that use multiplier chains to present the desired frequency to the final amplifier.

One would hope that there are very, very few spurs on a transmitter's signal, but one must be aware that they exist. The FCC requires amateur transmitters (a heavy source of EMI) with at least 5 watts of RF power to operate with at least 40 dBc (decibels below carrier) of suppression of unwanted signals. Standards for amateur transmitters operating at less than 5 watts are 30 dBc in the high-

frequency shortwave spectrum. Commercial standards are at least 60 dBc for all transmitters operated in the VHF/UHF region.

The transmitter must be "clean," i.e., operating within legal parameters, or it can be put off the air by the FCC. That puts a responsibility on the owner of the transmitter to operate in a clean manner. Once you are certain that the transmitter is operated in this manner, then you can troubleshoot the susceptible equipment to determine the cause of the problem.

THE ANTENNA

The antenna in a radio communications system is critical to EMI considerations. The antenna should be as far away as possible from the susceptible equipment—the higher and farther away, the better. Unfortunately, that is not always possible, for the antenna is in a fixed location. And quite often, that antenna location is set by the license of the station.

MODES OF ENTRY

There are three modes of entry for EMI into a system:

1. Radiation

2. Conduction

3. Magnetic induction

Radiation is direct signals from an antenna, through space, to the susceptible equipment. The receiver or other equipment is affected by the fundamental or a spur direct from the antenna.

Conduction arrives by the wires connected to the susceptible equipment. The wires include the AC power mains, the antenna (if any), other devices connected to the device, and ground leads (if any).

Magnetic induction means that the two circuits are magnetically coupled. A device such as a power transformer is nearby and causes problems. The degree of coupling depends on factors such as capacitance between the emitter and the susceptible equipment, the impedance levels involved, and the amplitude and frequency of the signals involved.

A special category is direct pickup of radiation. The circuitry of a piece of susceptible equipment picks up the signals on the printed circuitry and ordinary wiring. This type of EMI is particularly difficult to troubleshoot and repair because it involves shielding of the equipment.

DIFFERENTIAL-MODE VS COMMON-MODE SIGNALS

There are two modes of pickup of signals: differential mode and common mode. These two modes differ from each other, and the cures are different.

The differential-mode signal (Figure 3.2A) has two easily identified conductors, and the signal appears between them. The differential mode is characterized by two signals 180 degrees out of phase with each other, flowing in opposite directions on the wire.

The common mode (Figure 3.2B) occurs between two or more conductors in a multiconductor cable, and those cables act as if they were one cable. Common-mode signals are in phase with each other. It is usually (but not always) the case that common-mode signals involve a ground connection (Figure 3.2C).

The fixes for common-mode and differential-mode signals are fundamentally different. High-pass and low-pass filters are typical differential-mode fixes, while common-mode chokes (Figure 3.3) are typically used with that type of signal. The common-mode choke consists of a ferrite rod or toroid core (preferred) wrapped with the cable or cord in question.

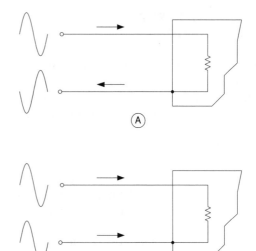
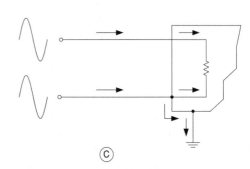

Fig. 3.2 *(A) Differential-mode signals; (B) common-mode signals; (C) common-mode signals with ground.*

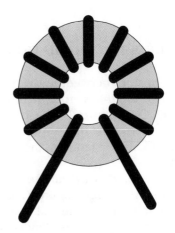

Fig. 3.3 *Common-mode choke.*

EQUIPMENT DESIGN CONSIDERATIONS

The design of the susceptible equipment is critical to the equipment's susceptibility to EMI. The EMC capability of the equipment is determined by things such as *filtering* and *shielding* of circuits.

The *filters* involved might be low-pass, high-pass, band-pass, or notch, depending on the frequencies involved. The filters may be in signal lines, or in the AC power mains if that is the entrance point being protected.

Shielding is metal or conductive surfaces placed over the affected circuits. It may take the form of a coating inside the plastic equipment cabinet. Shielding is fundamental to good design, but is often not done for economic reasons. Shields cost money, and they provide no benefit for the majority of consumers (those not subject to EMI).

Another factor is the design of the input amplifiers, especially if in integrated circuit form. The IC amplifier should be selected to have very nearly the same rise and fall times. Rise/fall time symmetry is important because of the capacitances that become charged during cyclic excursions of the input signal. If the rise and fall times are different, they tend to upset the input signal and manufacture differential signals where none existed before.

RADIO-FREQUENCY RADIATION

The relationship between power density and *field strength* is found throughout the EMI/EMC literature. Understanding the difference between these parameters is necessary to understanding the EMI/EMC problems.

The power density can be visualized by considering a huge sphere (Figure 3.4) radiating out from a spherical point ("A"). Imagine that the power is divided equally across the surface of the sphere. If we look

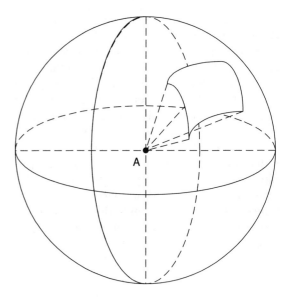

Fig. 3.4   *Sphere of electromagnetic signals.*

at the power on a square on the surface of the sphere that is 1 meter on a side, we get *watts per square meter* (watts/meter² or W/m²). The radiation falls off according to the *inverse square law* ($1/D^2$), which has implications for antenna placement with relation to EMI.

The field strength proceeds from the fact that an electromagnetic field has both magnetic and electrical fields. The electrical field is easiest to measure, so is used for the field strength measure. The electric field has maxima and minima, and the difference between them is the field strength. The units are volts per meter (V/m) or its subunits (mV/m or μV/m).

The power density and field strength are related to each other. The conversion is through the impedance of free space. This is determined to be 377 ohms per square meter (377 Ω/m²). We know that

$$V = \sqrt{P\,R}$$

from which we can calculate the voltage difference if $R = 377$. Consider an example. A 10-watt transmitter at a distance of 30 meters has a power density of:

$$P = \frac{10 \text{ watts}}{4\pi \, 30^2} = \frac{10}{11,309} = 8.84 \times 10^{-4} \text{ W / m}$$

The field strength is:

$$V = \sqrt{8.84 \times 10^{-4} \text{ watts} \times 377\Omega} = 0.577 \text{ volts}$$

NEAR FIELDS AND FAR FIELDS

The near field of a radiation source is that field where the component energies flow backwards and forwards between the radiating element (antenna) and the field in a manner that suggests that the inductance and capacitance of the antenna are being called into effect. The far field, or radiation field as it is sometimes called, is the field where the radiation behaves in a $1/D^2$ manner described above. The near field falls off at $1/D^4$.

The distance of the near field is open to some interpretation, but certainly by one-half wavelength the near field is extinguished. At the AM broadcast band, the near field can extend out quite a distance because of the low frequencies involved (540 to 1,700 kHz), but at VHF/UHF it extends only a few centimeters from the antenna.

Chapter 4

Grounding Methods
for RF Systems

Grounds and grounding are topics we hear a lot about in electronics and communications. The grounding concept is key to the proper operation of an electrical power system, to safety concerns, to radio and communications, and to reduction of electromagnetic interference. Despite its importance, many books on radio communications and electronics fail to address the topic in sufficient detail. In this chapter we will look at radio-frequency (RF) grounds, how to make them, and how to install them.

The word "ground" comes from the fact that it once referred specifically to a connection to the Earth. Although that connotation still exists, there are other "grounding" situations as well. The correct usage is that the word "ground" refers in American terminology to the common point in a circuit from which voltages are measured. In British terminology, a "ground" is referred to as "earth."

SCHEMATIC SYMBOLS

Figure 4.1 shows several symbols for grounds found in circuit diagrams. Although some of them tend to be used interchangeably, there are common practices that make them a bit different. For example, the ground in Figure 4.1A usually refers to an actual Earth ground connection, while the symbol in Figure 4.1B usually refers to the chassis ground on a piece of equipment. In some cases, you will see a wire leading from inside of a piece of equipment to outside that has both symbols to denote the fact that the chassis is earth grounded.

The symbol in Figure 4.1C is used to indicate a special ground. For example, it might be used for a low-signal-level ground, or it might indicate a ground (or more properly "common") connection that is isolated from the chassis or earth grounds. The versions shown in Figure 4.1D through Figure 4.1F are used in many foreign schematics,

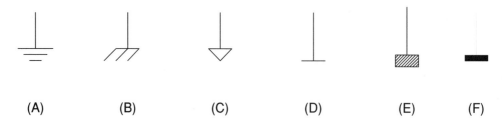

Fig. 4.1 *Ground symbols used in circuit diagrams.*

especially Japanese and European. In some cases, a "0V" is placed alongside the horizontal bar of the symbol to denote a "zero volt" condition.

DIFFERENT GROUNDS

Grounds are used for different purposes, and their usage determines in part their form—and also the standards to which they are built. There are electrical power system grounds, electrical safety grounds, lightning protection grounds, and RF-signal grounds.

In North America the standard electrical power system (Figure 4.2A) uses 60-Hz alternating current (AC). The power emerges from the power station at a high voltage and is transmitted over high-tension lines to your local substation. The substation transformer (T1) reduces the high tension to a lower voltage, but still at several kilovolts. This power is distributed over high-voltage lines to your neighborhood. A transformer (T2), often called a "pole pig" if it is on a telephone/power pole, reduces the several kilovolts AC to 240 volts center tapped.

There is a distinct economic advantage to the use of a single high-voltage line to distribute power over distances. The power level is the same despite the voltage. Because $P = V \times I$, the current will drop when the voltage increases, assuming power level is constant. Copper and aluminum wire are expensive. The use of high voltage means that smaller diameter (cheaper) wires can be used.

Figure 4.2B, and transformer T2 in Figure 4.2A, shows the standard electrical system. The secondary of transformer T2 (Figure 4.2A) is center-tapped. The center-tap is grounded and forms the neutral line (N). The ends of the transformer form two hot lines (H1 and H2). These three lines are fed into the house.

Figure 4.2B shows the standard wiring inside the residence. Two different voltages are available: 120V and 240V (these refer to RMS voltages, not peak voltages). A 120V outlet is made by using the neutral and one hot wire (e.g., either H1-N or H2-N) as the wiring. A 240V outlet is made by using both hot lines (H1-H2). In the case of the 120V outlet both the neutral (large hole in the socket) and the ground wires are carried back to the service box connection, where they are joined.

The principal ground at your home is the service entrance ground shown in Figure 4.2B. Other ground connections will be bonded to this point. The grounded scheme provides for safer operation in case of certain faults, but it also causes some hazards if you work on electrical equipment. Because the AC power in the building is ground referenced, touching either H1 or H2 while grounded can lead to painful and possibly fatal electrical shock. And what constitutes being "grounded"? Standing on a concrete floor with damp shoes or in your bare feet will do it.

Safety Grounds

Most electronic equipment has three wire electrical power cords. The third wire (colored green) is a ground wire and is kept separated from the neutral wire. Figure 4.3 shows the typical wiring of electronic equip-

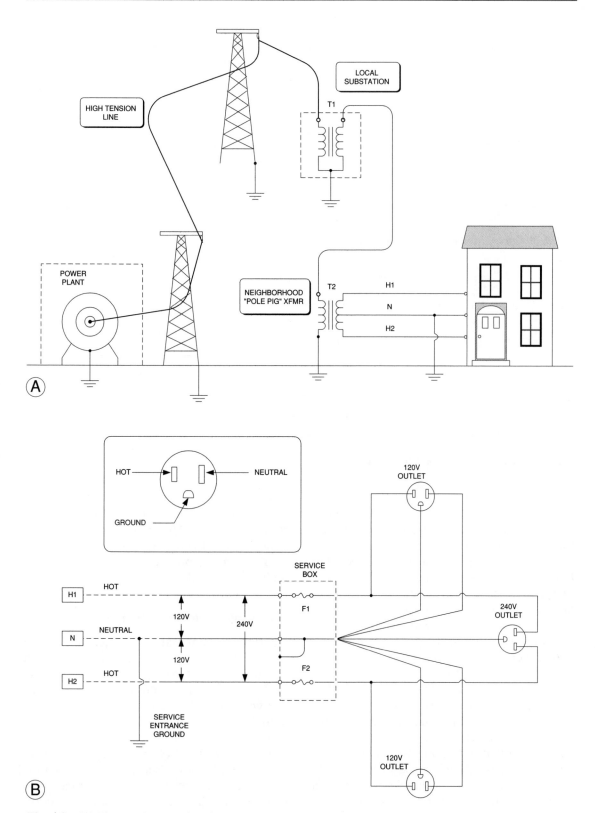

Fig. 4.2   *(A) Electrical power distribution system; (B) local wiring in your home.*

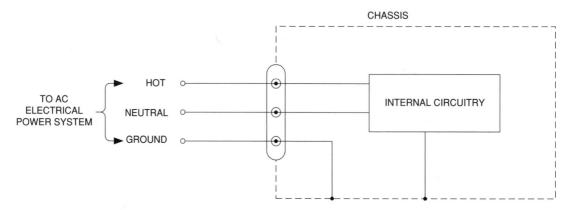

Fig. 4.3   *Safety ground on electronic equipment.*

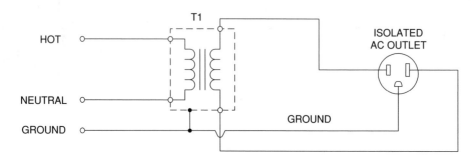

Fig. 4.4   *Isolation transformer power system.*

ment. The plug will contain the three wires (neutral, ground, and hot). The neutral and the hot lines are routed to the internal circuitry, while the ground wire is connected to the chassis or cabinet. The internal circuitry may also require a ground, so that line is also connected to the chassis. In some cases, a *counterpoise ground* wire is used in the circuitry. This "ground" is a common and may either float or be connected to chassis ground via a capacitor.

In some venues, the safety hazard is eliminated by using an isolation transformer (T1 in Figure 4.4). Hospital operating rooms are often wired with a 10- to 30-kVA isolation transformer that serves only that room. If you are unfortunate enough to be inside an OR, then you may notice a large stainless steel panel with a number of green and red lamps. It may also have a clock. Inside that remov-

able panel is an isolation transformer. The reason is safety. My own electronics workbench is wired with a 1.6-kVA isolation transformer. Professional electronic servicers (if they are smart) do the same.

Lightning Protection Grounds

Lightning is a massive electrical discharge from cloud-to-cloud or cloud-to-ground. In the latter case, if you or your property gets in the way of the lightning bolt, then damage will occur. Tall structures, including antennas and antenna towers, are grounded as protection against lightning strikes. The grounding will not eliminate damage, but it will substantially limit it. If you put up a radio antenna tower, the local electrical code will require grounding . . . and for good reason!

RF Signal Grounds

Radio-frequency systems (receivers and transmitters) are usually grounded for both safety and performance reasons. The ground is not strictly necessary for radio communications, especially if a well-balanced antenna is used. But other forms of antenna (random length wire Marconi and vertical) won't work without a "good ground." Even when the system uses a balanced antenna (e.g., dipole with BALUN), a ground should be provided. This prevents certain types of interference, and also eliminates problems of interaction with other pieces of equipment in the radio station.

Unfortunately, many books on radio or radio antennas don't go into a lot of detail as to what a "good ground" is, or how to get it.

GROUNDING SYSTEMS

A good ground has a low resistance path to earth. How it gets there depends on how well you design and install the ground system. Figure 4.5 shows a typical setup for a radio station (amateur, broadcasting, or two-way). There are several pieces of equipment mounted on a desktop, or in a cabinet or rack of some sort. The AC power strip must be a three-wire type. The ground bus can be a copper plate, copper roofing "flashing," or a copper pipe of 0.5- to 2-inch diameter. The chassis of the equipment are connected to the ground bus through heavy wire or wire braid. In the case where a copper pipe is used for the ground bus, hose clamps can be used to connect the equipment ground wires to the bus.

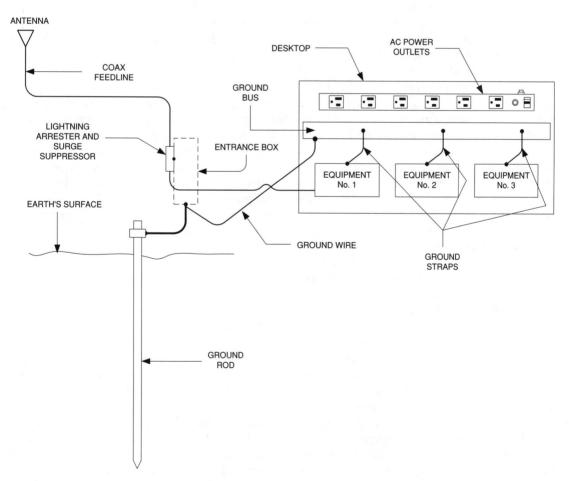

Fig. 4.5 *Ground rod and its connections.*

The ground bus is connected to the external ground system that is actually earthed through a heavy ground wire. The ground system should be a ground rod at least 8 feet long—but more about that in a moment.

In the case of a radio system, the antenna feedline is a source of potential danger. If lightning strikes the antenna or anywhere near it, then a real problem is created! As a result, proper antenna installations use an approved lightning arrester. These devices (ideally) divert a large part of the energy from the lightning to ground, although it is likely that some damage will still exist. Some lightning arresters also include surge suppressors of the sort used for computer AC power lines. These devices will remove voltage transients that can damage the electronic equipment, but are less than lightning strikes. The lightning arresters are mounted outside the building. A metal entrance box or panel made of metal is used to hold the devices.

It is commonly believed that cold water pipes can be used for grounding radio and electronic equipment. That advice held true at one time, but is not good today. Several problems are found. First and foremost is the fact that plumbing pipes are not made of metal anymore—they are of PVC or other synthetic materials. In other words, insulators. In other cases (and this was true when pipes were commonly used for grounding), the electrical properties of the ground are defeated by the use of thread compounds to seal the system. Pipe threads leak water unless covered with pipe thread compound, or more commonly today, *Teflon* tape. There are other good reasons to not use the cold water pipes, but these are the principal reasons.

Also, don't depend on the AC power line ground wire for radios or sensitive electronic equipment. I've seen radio transmitters act as if they were ungrounded because the owner relied on the "green wire" inside the power cords. In another case, electronic instruments used in a medical school were disrupted occasionally by problems that we eventually traced to noise signals on the ground line. A properly designed scientific facility would have instrument ground systems separated from the electrical power system (just like radios).

Ground Rods

The short ground rods sold for TV antennas are not suitable for proper ground systems. The minimum ground rod will be an 8-foot copper or copper-clad steel rod either 5/8-inch or 3/4-inch in diameter. Alternatively, a copper pipe can be used. The ground rod or pipe should be driven into the soil until only 4 to 6 inches remains above grade. Clamps are used to connect the ground wire to the ground rod, not solder (solder will melt in the case of a lightning strike or heavy AC current fault).

An AM/FM radio station I know uses a 4-inch diameter, 16-foot long thick-wall copper pipe. It was planted in the ground by drilling a well hole, and then the transmitter building was constructed over it. The pipe passes right through the bottom of the FM transmitter.

Figure 4.6 shows detail for the ground rod connection. Several different forms of wire clamp are available. In some designs the ground wire is clamped beneath the screw that holds the clamp fast to the rod. Those types are equipped with a flat piece of metal on the end of the screw. In other cases (perhaps most commonly) the wire is clamped to the pipe opposite the screw. When the screw is tightened it cinches the clamp against the ground rod, making electrical contact. Finally, some people use hose clamps to fasten the wire to the rod. What should not be used, however, are solder connections.

GROUND DESIGNS

The common AC power ground may be a single 8-foot ground rod driven into the earth. Radio stations, especially high-power stations, tend to use more complex grounds. Whatever the case, the goal is to provide a ground with a resistance of 10 ohms or less. An electrician friend told me that our local electrical code requires extra grounding if the ground resistance is 25 ohms or more.

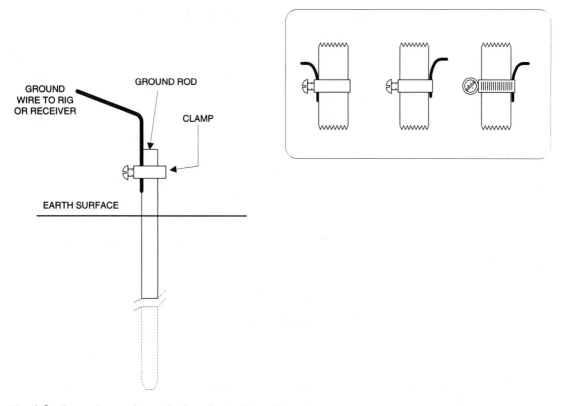

Fig. 4.6 *Grounding radio and other electronic equipment.*

The factors that affect ground resistance include the soil resistivity, the resistance of the ground rod, and the contact resistance of the rod and soil.

Soil Resistivity

The soil resistivity varies markedly with the type of soil. Table 4.1 shows several different types of soil composition and the range of resistivities (in ohm-cm) expected. The actual value depends on local composition, moisture content, and even soil temperature. The resistivity at any one site may tend to vary over the course of time as these factors change.

People working with radio systems and antennas often use descriptions such as those shown in Table 4.2 for soil resistivities. These relate the type of terrain or development of the area to a range of resistivity values. If you do antenna modeling, you will find that the nature of the soil forming the ground plane beneath the antenna profoundly affects performance.

Table 4.1   Resistivities of Various Soils

Type of Soil	Ohm-cm
Surface soil (loam, gumbo)	10 to 5K
Clay	200 to 10K
Sand and gravel	5 to 100K
Surface limestone	10K to 1 meg
Deep limestone	500 to 400K
Shale	500 to 10K
Sandstone (dry)	10K to >1 meg
Sandstone (wet)	200 to 10K
Granite	100K
Basalt	100K
Decomposed gneisses	5 to 50K
Slate	1 to 10K
Fresh water (lakes)	20K and up
Tap water	1 to 50K
Sea water	20 to 200K

Table 4.2   Resistivities of Various Types of Terrain

Type of Terrain	Ohm-cm
Pastoral, low hills, rich soil	300 to 1K
Flat country, rich soil	100 to 10K
Densely wooded	100 to 3K
Marshy	100 to 1K
Pastoral, medium hills	300 to 10K
Rocky soil (e.g., New England)	1 to 100K
Sandy, dry coastal areas	30 to 500K
City or industrial	100K to 1meg
Fill land	600 to 70K
Suburban (mixed composition)	300 to 20K

Ground Rod Resistance

The resistance of the ground rod is a function of the resistivity of the material it is made of, the resistance of the wires, and the resistance of the clamps. If there is a problem with ground rod resistance, then look first at the clamps and wires. A ground rod in good condition will have a negligible resistance, so any problem is bound to be in the wire or clamp.

Contact Resistance

The resistance of the rod in contact with the soil depends on the length of the rod buried in the soil, and on how firmly the soil is packed around it. It also assumes that the

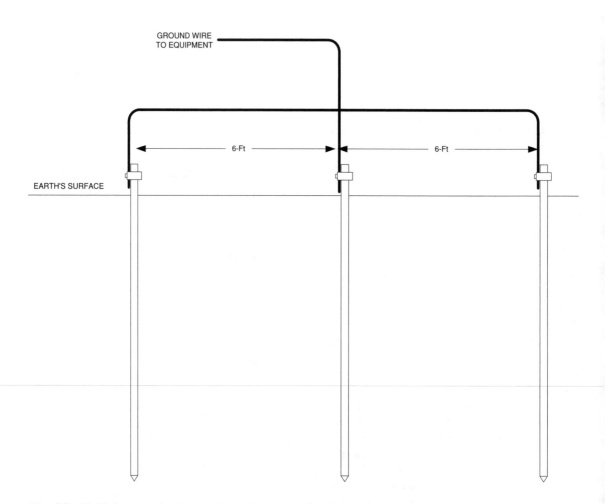

Fig. 4.7   Multiple ground rod system to reduce ground resistance.

ground rod was unpainted, unvarnished, and free of surface grease and other contaminants.

Ground rod contact resistance also depends on the diameter of the ground rod, but that effect is limited. You can expect a 20 to 30 percent reduction when the ground rod is increased from 0.50-inch to 1.0-inch, but when the diameter is doubled again to 2 inches only a 5 to 8 percent reduction will be seen. Although 0.50-inch and 1-inch ground rods are used, the usual compromise is to use 5/8-inch or 3/4-inch ground rods.

Multiple Ground Rod Systems

One of the most practical ways to reduce ground resistance is to use multiple ground rods spaced at least 6 feet apart (Figure 4.7).

In some cases, the ground rods are laid out in a straight line, but in others a pattern such as a circle, square, rectangle, or triangle is used.

Figure 4.8 shows the reduction in resistance as a function of rod length for systems of one, two, three, and four ground rods. If you are forced to use shallow ground rods, then note that the reduction in resistance is very profound as the number of rods is increased. For longer (i.e., deeper) rods, however, increasing above a certain number reaches a point of diminishing returns.

OTHER GROUND ELECTRODES

The ground rod is only one form of electrode used to connect equipment to the earth.

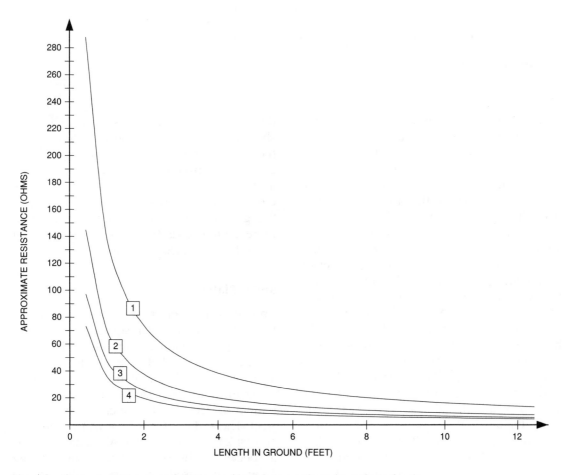

Fig. 4.8   Ground resistance as functions of length in earth and number of rods.

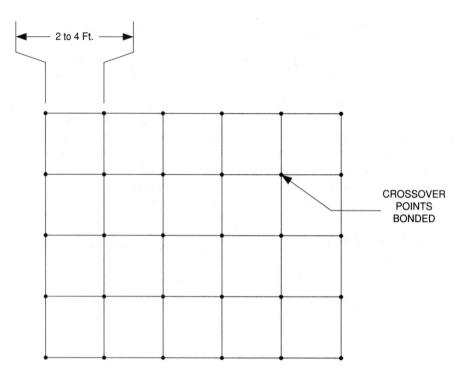

Fig. 4.9 Ground wire grid.

Early in my career I knew a rather eccentric amateur radio operator named Abe. When his house was being built in 1940 he took care of his ground in a classy manner. When the contractor dug the basement, they also dug a hole alongside the house, a few feet from the house. Abe dropped an old copper bathtub into the hole, and then backfilled with rich soil. A heavy set of copper wires had been brazed to the copper tub before backfilling. Those wires were routed through the wall of the basement to a large copper plate at one end of Abe's ham station.

But ol' Abe wasn't finished. After the house was built, and the only thing left to do was put the sod in place, he and several ham buddies went out to the site and laid down a grid of #10 bare copper wires all over the lot. The following day the sod contractor laid down the lawn.

AM radio stations often have grounding systems like Abe's. The use of antenna radials and buried grids is quite common. The ra-

dials will be discussed shortly. The grid should ideally be buried at least 6 inches in the ground, with the wires making up the grid being spaced 2 to 4 feet apart (Figure 4.9). The wires are bonded together at the crossover points. In some cases, the wires of the grid will also be connected to one or more ground rods (or presumably a copper bathtub in the case of ol' Abe).

Metal Plates

Another form of electrode is the buried metal plate electrode. Circular, square, or rectangular copper, aluminum, or steel plates are buried 5 to 10 feet below grade (Figure 4.10). The plates should be at least 60 mils (0.06 inch) thick in the case of copper and aluminum or 250 mils (0.25 inch) in the case of steel. The surface area of the plate should be at least 2 ft^2. I suppose that Abe's bathtub was a species of the metal plate ground. Other "Abes" have used automobile radiators and

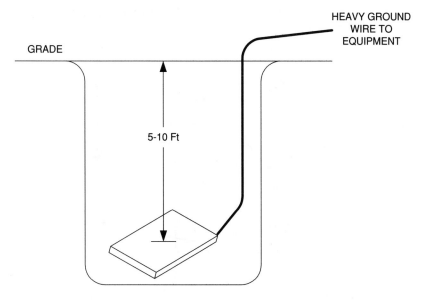

Fig. 4.10 *Buried plate electrode.*

engine blocks for ground electrodes with good results.

Horizontal Conductors

In cases where bedrock is near the surface, or there is a layer of hard material such as marine clay or limestone aggregate close to the surface, a set of buried horizontal conductors can be laid out in row (Figure 4.11). The conductors could be regular ground rods laid on their sides, copper plumbing pipe, aluminum tubing, heavy solid or stranded wires, or strips of roofing flashing. The conductors are buried in slit trenches 18 to 40 inches deep, or whatever it takes to get below the local "freeze line."

Electrolytic Grounding Systems

Environmentalists are going to love this type of grounding scheme. Electrolytic ground systems use either periodic or continuous alteration of the soil resistivity by the addition of salts. Forms of soluble salts used for ground resistivity improvement include magnesium sulfate (Epsom salts), copper sulfate ("blue vitriol"), calcium chloride, potassium chloride, and sodium chloride. Although powered or granulated salts are used, the solid "rock salt" forms are preferred.

CAUTION
Keep in mind that these salts can poison nearby vegetation, and may also interact with the concrete or cement used in building construction.

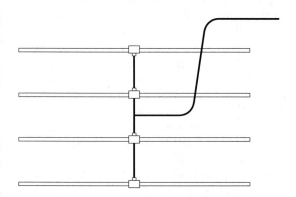

Fig. 4.11 *Buried horizontal conductors.*

Slit and doughnut trenches are sometimes used to make an electrolytic ground

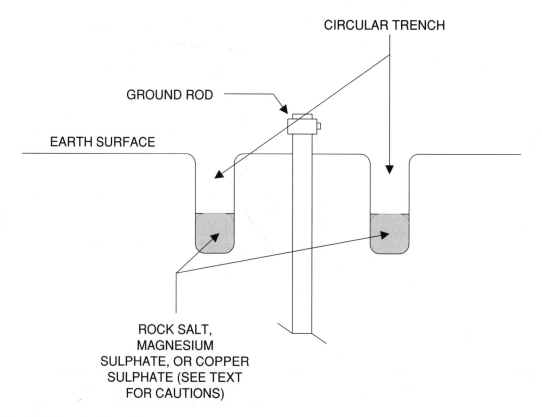

CIRCULAR TRENCH

GROUND ROD

EARTH SURFACE

ROCK SALT,
MAGNESIUM
SULPHATE, OR COPPER
SULPHATE (SEE TEXT
FOR CAUTIONS)

Fig. 4.12   Doughnut trench.

system. In the case of a slit trench, the ground rod would be installed in the center of the trench, pointed end down. The trench is typically 12 to 24 inches deep, and about 6 to 12 inches wide.

The doughnut trench depicted in Figure 4.12 surrounds the ground rod. It is preferred for longer term installations because it keeps the corrosive salts away from the ground rod and its fittings.

For short-term installations the trenches can be filled with salt water. Although sea water is sometimes used, a solution of 1-lb salt per gallon of water is better. Prepare enough salt water to fill the trench. Longer term installations should be done with rock salt or other coarse forms. The salt is then placed in the trench, and an overburden of gravel and soil is backfilled into the trench. Rainwater (or your garden hose) will dis-

solve the salt. Treatment every 4 to 6 weeks is necessary to keep the soil resistance low in rainy climates, but others treat annually.

An electrolytic "salt pipe" scheme is shown in Figure 4.13. The pipe holds one of the salts mentioned above. The pipe can be made of copper, steel, or even PVC (it need not be conductive). A large number of 0.25- to 0.375-inch holes are drilled into the surface of the pipe, and end caps are fitted onto the ends (some versions don't use the bottom end cap). The pipe is buried 18 to 48 inches into the soil, with only the top appearing above grade. Once the pipe is in place it is filled with rock salts or other solid forms. When the soil moistens, as during rainstorms, salt will leach out of the pipe into the surrounding soil. It usually takes a number of these pipes around an antenna such as a vertical to do a lot of good.

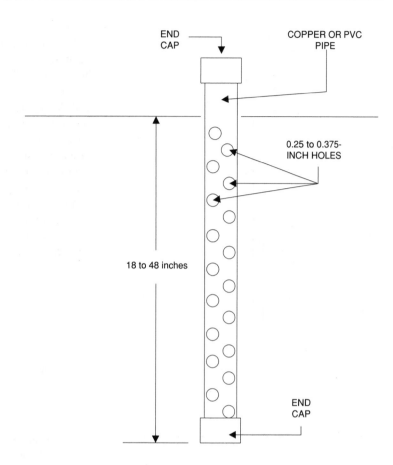

END
CAP

COPPER OR PVC
PIPE

0.25 to 0.375-
INCH HOLES

18 to 48 inches

END
CAP

Fig. 4.13
Electrolytic salt pipe.

INSTALLING GROUND RODS

Assume that you've considered plates, bathtubs, and electrolytic ground systems, but finally, after much engineering consideration, decide to install a simple 8-foot copper-clad steel ground rod. Sounds easy. And sometimes it is . . . but don't count on it.

The traditional brute-force approach to installing ground rods is shown in Figure 4.14. The rod is placed pointed end down on the top of the soil, and a sledgehammer is used to apply repeated blows to the upper end of the rod. It's a lot of "fun" standing on a stepladder swinging a 3-lb sledgehammer! In some areas the rod will slip right into the ground. But in others, the progress is remarkably slow. In some cases, progress comes to a halt rather quickly.

I owned one house where my amateur radio ground rod was nearly impossible to install. About a foot below grade there was a 2-foot deep layer of gray marine clay. That stuff could not be penetrated by brute force methods. Indeed, repeated blows to the top end of the ground rod would flare the end of the rod and prevent installing the wire clamp later on. There had to be a better way.

Better Ways

Several people have related ways of installing ground rods that are more easily done. One method is to get a "well point" tool and drive it into the soil to create a channel for the ground rod. Another method is shown in Figure 4.15. A posthole digger is used to create a round hole about 6 inches in

diameter. The ground rod is tapped into the bottom of the hole, and then supported upright for a while. Water is poured into the hole up to the top. Leave the water alone, and in 20 to 90 minutes it will seep into the soil. When the water is largely gone, use the sledgehammer to drive the point of the ground rod further into the ground, until significant resistance is noted. Fill the hole with water again, and wait until it seeps down. This process might take a couple of days in some particularly difficult areas, but it works.

One final method is shown in Figure 4.16. This method works nicely for really difficult cases. The "ground rod" is a copper pipe

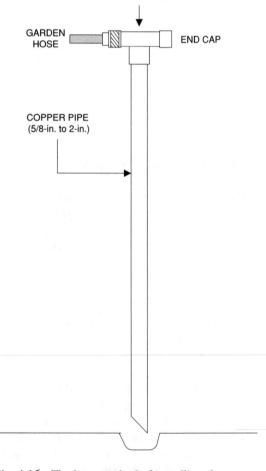

Fig. 4.14 *Traditional way of installing ground rod.*

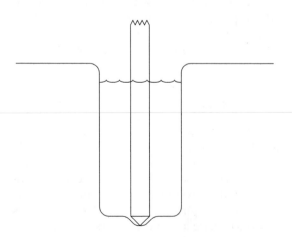

Fig. 4.15 *An alternative method.*

Fig. 4.16 *The hose method of installing the ground.*

5/8 inch to 2 inches in diameter. The bottom end of the pipe is beveled to make it a sharp point. The top end of the pipe is fitted with a "tee" connector. One end of the tee is fitted with an end cap sweat soldered in place, while the other end is fitted with a hose nozzle fitting. That connection allows a garden hose to be connected to the ground rod.

By driving water into the pipe from the garden hose, and using the "Tee" as a handle to apply downward force, you'll be able to slip the pipe easily into the ground. Once the pipe is in place, the tee can be cut off with a hacksaw, and a wire clamp fitted.

TOWER GROUNDING

Antenna towers must be grounded for lightning protection. Typical towers used to support antennas are built in a triangular footprint pattern (i.e., are three-sided) and rise 30 to 150-feet in height. The tower is mechanically mounted to a base plate that is fitted to a concrete pedestal buried in the ground. A copper ground ring made of pipe or heavy wire circles the tower base (Figure 4.17). At points around the copper ring 8-foot ground rods are driven into the soil and bonded to the ring. Heavy braid or ground wire is used to bond the ground rods to the metal legs of the tower.

The typical antenna mounted on this type of tower is fed with coaxial cable. In order to keep voltage fields from causing damage in the event of a strike, it is common practice to ground the coax to the tower at both the top and the bottom of the tower.

VERTICAL ANTENNA COUNTERPOISE GROUNDS ("RADIALS")

Marconi antennas are unbalanced with respect to ground (Hertzian antennas like the

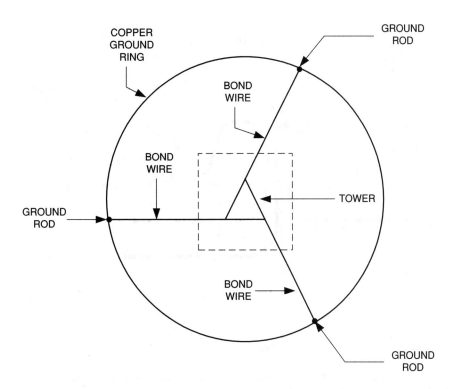

Fig. 4.17 *Grounding an antenna tower.*

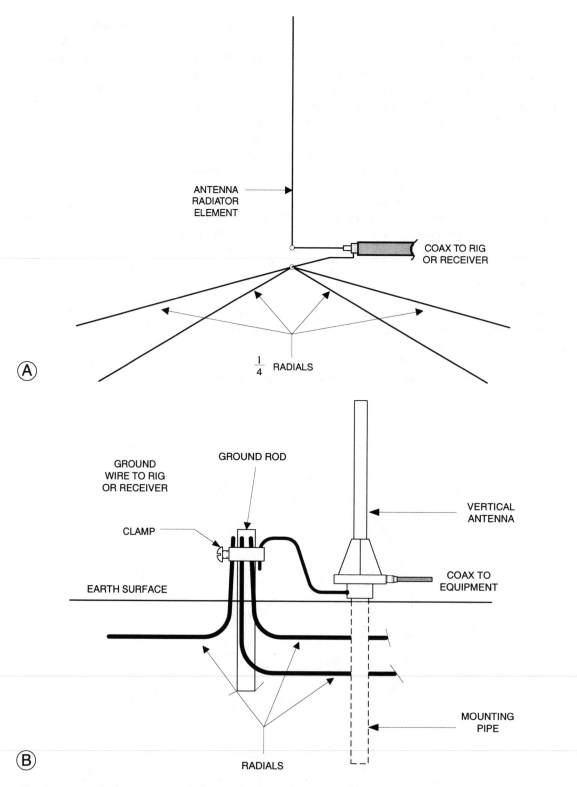

Fig. 4.18 *(A) Vertical antenna with radials as counterpoise ground; (B) ground connection for buried radials and ground rod; (C) relative ground resistance as a function of the number of buried radials.*

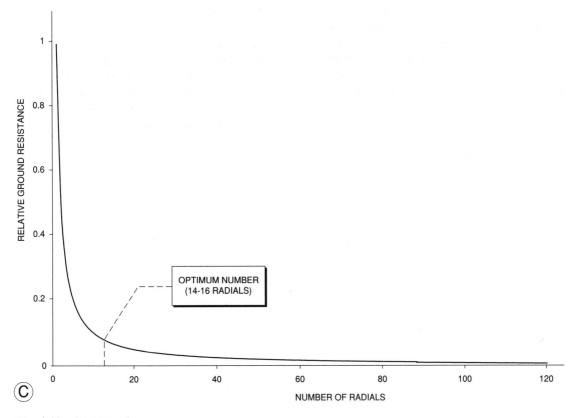

Fig. 4.18 *(continued)*

dipole are balanced). Included in the Marconi class of antennas are random-length wires, certain long wires, and vertical antennas. The latter form come in sizes from 1/4-wavelength to multiple wavelengths. One side of the transmitter output circuit must be grounded for this antenna to work properly. Figure 4.18A shows a typical vertical antenna. The radiator element is connected to the transmitter through the center conductor of the coaxial cable feedline. The shield (outer conductor) of the coaxial cable is connected to ground. But what if the antenna is mounted above ground? In that case we use a counterpoise ground consisting of quarter-wavelength radials.

Some people claim that the worst problems with ground-mounted vertical antenna performance are due to poor grounding. As a result, it is common to see a vertical antenna mounted close to the ground fitted with a

set of radials. Typically, for safety's sake, the radials are grounded (surface installed radials are a pedestrian hazard). Figure 4.18B shows how the radials are installed. Sometimes, they are paired with one or more 8-foot ground rods. The radials are buried 6 to 12 inches below grade and are made of bare copper wire. The length is not critical, but most authorities claim that they should be at least quarter-wavelength. However, others will tell you that any radials are better than none.

The number of radials determines how well the system works. But there is a limit to how much goodness makes sense. By standard, AM broadcast stations in the United States have 120 radials. But amateur radio and some other sources claim that 14 to 16 radials are a practical optimum number (Figure 4.18C) because of diminishing returns above that number. Figure 4.18C shows the relative ground resistance as a function of the number of radials buried in the soil. There is a relatively sharp knee between 15 and 20 radials, beyond which the ground resistance drops much less for each additional radial. Given the cost of burying radials, the practical limit seems to be about 14 to 16 radials.

CONCLUSION

A "good ground" is necessary for the proper functioning of certain antennas, some types of radio systems, other electronic equipment, and the AC power systems of homes. In this chapter you have seen a number of aspects of the grounding problem.

Chapter 5

Shielding Electronic Circuits

It is almost an article of religion in electronics that shielding electronic circuits prevents EMI problems. A good shield will keep undesired signals inside in the case of a transmitter, or outside in the case of other forms of circuits. All transmitters generate harmonics and other spurious signals. If these signals are radiated, they will interfere with other services. Signals that go out through the antenna terminal usually pass through either tuning or filtering networks, which tend to clean up the emission. But if the circuits are not shielded, direct radiation from the chassis will defeat the effects of the filtering.

Shields are good. Unfortunately, many shields are essentially useless. In some cases, they may even cause more problems than they cure. The problem is not just on transmitters, or even just on RF circuits in general, but on all electronic circuits. I once worked with medical and scientific electronic instruments that rarely used frequencies above 1,000 Hz, and they were subject to severe EMI. Why? The 60-Hz power line EMI!

Let's look at shielding materials and methods. Figure 5.1A shows a universal "black box" circuit with three ports: "A" is the input, "B" is the output, and "C" is the common. The term "black box" means any form of electronic circuit . . . it is used to universalize the discussion so that ideas are not associated with any specific class of circuit. What's inside the black box enclosure could be a circuit, or a system such as a transmitter, receiver, audio amplifier, or a medical electrocardiograph amplifier. It doesn't matter for our present purposes.

Shielding means placing a metal screen or barrier around the circuit. In Figure 5.1B the black box circuit is placed inside of a shielded compartment, as indicated by the dotted lines. In addition, the input voltage (V_{AC}) and output voltage (V_{BC}) are shown (the subscript letters refer to the port designations). Any time two conductors are brought into close proximity to each other, and not touching, a capacitance will exist between them. Sometimes the capacitances are intentional. But in other cases, the capacitances are incidental to construction (e.g., an insulated wire lying on a chassis creates a capacitance). In the case of Figure 5.1B there are three "stray" capacitances represented: C_{AD}, C_{BD}, and C_{CD}.

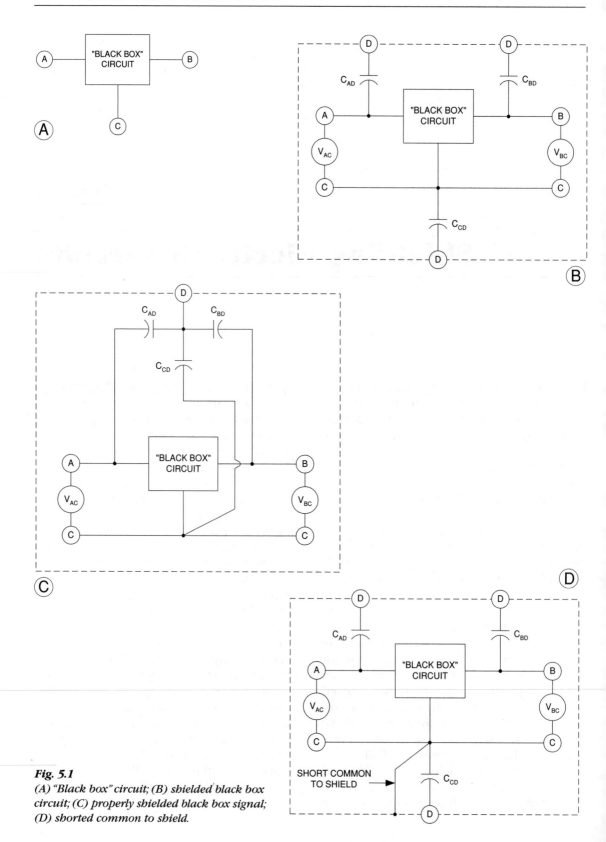

Fig. 5.1
*(A) "Black box" circuit; (B) shielded black box
circuit; (C) properly shielded black box signal;
(D) shorted common to shield.*

Shielding of the sort shown in Figure 5.1B is not terribly effective, and under some circumstances it can lead to instability and outright oscillation. The feedback path that causes the problem is better seen in the redrawn version of the circuit shown in Figure 5.1C. Capacitors C_{BD} and C_{CD} form a capacitive voltage divider, with the "output" connected through C_{AD} to the input terminal ("A") of the black box. Under the right circumstances, this circuit can lead to very bad EMI/EMC consequences.

The solution to the problem is to apply Shielding Rule No. 1:

RULE OF THUMB: The shield must be connected to the zero signal reference point in the circuit being protected, i.e., the common line between output and input.

In some cases, the common might be a floating connection that is not earth grounded, i.e., a counterpoise ground plane. The common point may be at a nonzero voltage, but for the purposes of the input and output signals it is the zero-signal reference point. In most cases, the zero-signal point is, in fact, at a potential of zero volts.

The application of this rule is shown in Figure 5.1D. The common ("C") is connected to the shield ("D"), effectively shorting out capacitance C_{CD} and the common node of the feedback network evident in Figure 5.1C. To restate the general rule: connect the shield and common signal line together.

In the case of Figure 5.1 the "black box" circuit is single-ended, so the common line of the internal circuit is connected directly to the shield.

Figure 5.2A shows a situation that is a little more complex. In this case, some "black box" circuit is inside of a shielded enclosure and supplies output signal to some sort of resistive load. The load is connected to the shielded enclosure by some sort of shielded cable. Similarly, a shielded signal source V_{in} is connected to the input of the "black box" by another length of shielded cable.

In this case, there could be too many grounds. Suppose that the common signal point inside the main shielded compartment is connected to the shield, and the shield is, in turn, grounded ("A"). The signal source is also grounded ("B"), but to a different point. If a current (I_G) flows in the ground plane resistance (R_G), then a voltage drop V_G will be formed across the resistance of the ground path. The current might be due to external circuits, or it may be due to a potential difference existing between two points in the circuitry inside the shielding. Whatever the source, however, a difference of potential between points "A" and "B" gives rise to a spurious signal voltage V_G that is effectively in series with the actual signal voltage (V_{IN}). Thus, the total signal seen as valid by the circuit is $V = V_{IN} + V_G$. This is the "ground loop" problem.

The key to solving the ground loop problem is to connect the shield to the ground plane at the signal end ("B"), and not at any other points. An application of Rule No. 1 might say: "The shield and common of the internal circuitry should be connected together at the point where the signal source is grounded." In other words, break the connection at point "A" and rely instead on point "B," as shown in Figure 5.2B.

This class of problem is representative of a class of problems in which a common impedance (in this case a resistance, R_G) couples two segments of a circuit. If a voltage drop appears across the common impedance, then a problem will surface.

APPROACHES TO SHIELDING

There are two basic approaches to shielding: absorption and reflection. These mechanisms often operate together. Suppose a large external field is present. In the case of absorption, the field may penetrate the shield but is greatly attenuated. In the case of reflection, the field is turned back by the metal shield. The absorptive method is usually used at frequencies below 1,000 kHz for magnetic fields. The types of materials tend to be the ferromagnetic materials such as steel and a special material used especially for magnetic shields called "mu-metal" or μ-metal. At higher frequencies, especially

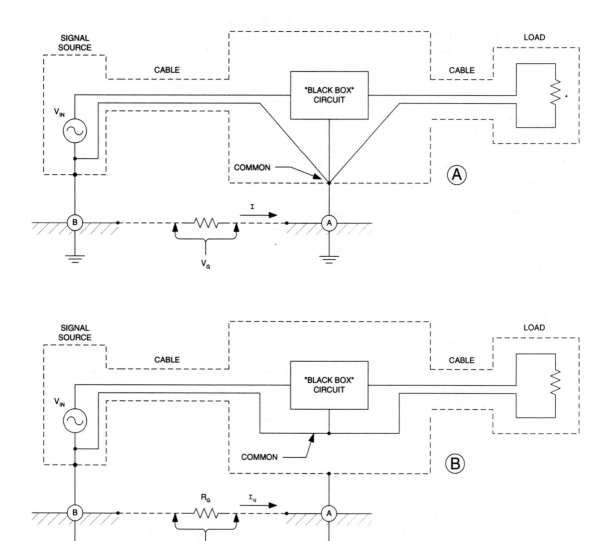

Fig. 5.2 *(A) Improperly grounded black box circuit; (B) properly grounded circuit.*

where the electric field is of more importance than the magnetic field, better shielding materials are copper, brass, and aluminum.

SKIN EFFECT AND SKIN DEPTH

Alternating currents (AC) do not flow uniformly throughout the cross-section of a conductor as is the case with direct currents (DC). Because of the skin effect, AC flows only near the surface of the conductor. This effect creates a situation where the AC resistance of a conductor will be higher than the DC resistance. If we graph the current density from the surface to the center of a cylindrical conductor, we'll find that the curve is a section of a parabola. The *critical depth* for a cylindrical conductor is the depth at which the current density falls off to 0.368 times the surface current density. It is this current that we use to determine the AC resistance.

Sheets or plates of metal used for shielding also show a skin effect when currents flow in them. The skin depth (Figure 5.3) is analogous to the critical depth in cylindrical conductors. In both cases, 63.2 percent of the current flows in the area between the surface and the skin depth (δ). The skin depth is calculated from:

$$\delta = \frac{2.602\,k}{\sqrt{F_{HZ}}} \qquad (5.1)$$

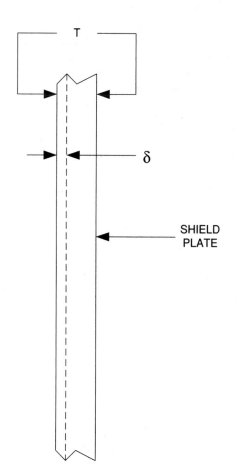

Fig. 5.3   Shield plate.

T

δ

SHIELD
PLATE

where δ is the skin depth in inches (in.), F_{Hz} is the frequency in hertz (Hz), and k is 1.00 for copper, 1.234 for aluminum, and 0.100 for steel.

Why is this important? In the case of absorptive loss, the attenuation is 8.7 dB/δ. For example, at 60 Hz a steel shield has a skin depth of 0.034 inches. If 1/16-inch stock is used, the total depth is equivalent to 1.84δ, so the attenuation for magnetic fields would be 8.7 dB × 1.84 = 16 dB.

To obtain maximum reflective loss at RF frequencies, the thickness of the shielding material should be about 3 to 10 times the skin depth (the thicker the shield, the better the shielding, up to a point). For example, at 10 MHz aluminum has a skin depth of 0.001 inches, and copper has a skin depth of 0.0008 inches, so the shield thickness should be 0.010 inches for aluminum and 0.008 inches for copper or more. Given that 1/16-inch thick stock is 0.0625 inches thick, aluminum will be a bit marginal, while copper would be more than sufficient. It is only fair to note, however, that some textbooks say a shield should be at least three times the skin depth . . . but that is for minimal shielding.

GROUND PLANES

The ground plane might be an actual earth ground, but in most electronics circuits it will be either a printed circuit board or a chassis. In the case of printed circuit boards it is usually recommended in RF circuits to use a double-sided board with the top side copper used as a ground plane, and possibly to carry DC power supply lines.

In RF circuits it is not advisable to use small wires or printed circuit tracks as ground lines. The AC resistance of cylindrical wire conductors is a function of both the wire diameter and the frequency. For any given wire size, the AC resistance:

$$R_{AC} = kR_{DC}\sqrt{F_{MHz}} \qquad (5.2)$$

Table 5.1   Wire Size K-Factor

Wire Size (AWG)	K-Factor
8	35
10	28
14	18
18	11
22	7

The value of the K factor depends on the wire size, as shown in Table 5.1.

Thus, when you use #22 AWG solid hookup wire to carry a 1-MHz RF current, the AC resistance is seven times the DC resistance. If this wire is a ground and carries a current, the AC resistance of the wire might be considerable, creating a nasty ground-loop voltage drop.

Even if the wire is large enough to reduce the effects of AC resistance at RF frequencies, the inductance might be a problem. The inductance of a straight length of #22 AWG wire is about 600 μH per 1,000 ft. A 1-ft run of wire will, therefore, have an inductance about 0.6 μH. This inductance will not be noticed in an audio circuit, or even many low-frequency RF circuits, but as the frequency climbs it becomes significant. In the upper HF and lower VHF regions it is a significant portion of lumped inductances intentionally placed in the circuit.

If the wire is in a ground path, then it is a common impedance. Any RF voltage developed across its inductive reactance forms a valid signal and may cause problems. The key to the problem is star grounding, i.e., grounding all circuit elements to the same point. If the signal source is grounded, then its ground connection ought to be used as the overall grounding point.

SHIELDED BOXES

A number of manufacturers sell prefabricated shielded boxes. Some of them are quite good, while others are not very good at all. Figure 5.4 shows one of the poorer forms of aluminum shielded box from an EMI point of view. It consists of two half-shells. The bottom shell is bent into a "U-channel" shape (see end view). The top shell is slightly larger and is designed to fit over the bottom shell. A pair of tabs on each side of the top shell either overlap or fit into mating notches in the bottom shell. This type of box is suitable for low-frequency (up to a few kilohertz) and DC applications that are not particularly sensitive to external EMI.

A somewhat better form of box is shown in Figure 5.5. The bottom shell is essentially the same as in the other box, but the top shell is built using an overlapping lip rather than

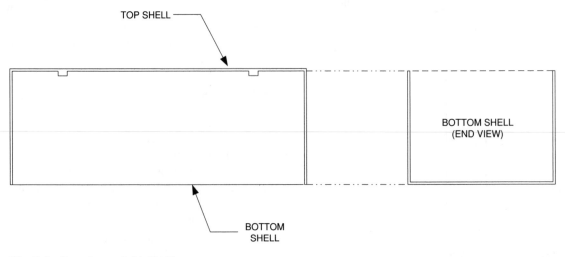

TOP SHELL

BOTTOM SHELL
(END VIEW)

BOTTOM SHELL

Fig. 5.4   Poor form of shielded box.

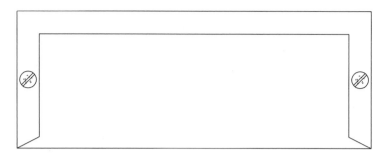

Fig. 5.5
Better form of shielded box.

tab-and-notch construction. This form of construction is good up to several megahertz, but may fail in the VHF-and-up region.

HOLES IN SHIELDS

Ideally a shield should contain no holes, but in practice this is impossible. There are always some connections (input, output, DC power) that must go in or out of the shielded enclosure. In other cases, the circuitry may generate considerable heat so some holes are provided to ventilate the interior. The holes must be very small compared to the wavelength of the highest frequency signal being protected against.

The general rule is that screw or mounting holes should be spaced not more than 1/20 wavelength (i.e., 0.05λ) apart at the highest frequency of operation. At 1 MHz, this is not hard to meet, because 0.05λ is more than 49 feet. But at VHF and up it might be a bit tricky because the wavelengths are

much shorter. For example, spacing the screws that keep a shield firmly in place 3 inches apart may be sufficient for mechanical strength and will shield at lower frequencies. But 3 inches is 0.05λ at 197 MHz. Above 197 MHz the shielding effect is therefore reduced. In the shielded enclosure of Figure 5.5 the screws are on the end portions of the flange. At UHF frequencies this can be a problem. A better solution is shown in Figure 5.6, where spacing S < 0.05λ.

The effects of wide spacing of mounting screws can be dramatic. I once saw a case where a mechanical engineer had "redesigned" the specification for an RF enclosure because she didn't understand the RF effects. But the electrical engineer designing the box showed her by taking a well-shielded pulsed RF transmitter and connecting it to a dummy load. He then used a spectrum analyzer with a whip antenna on it to monitor the energy emitted from the RF box. He started by removing every other screw. As soon as the first screw was loosened the harmonics

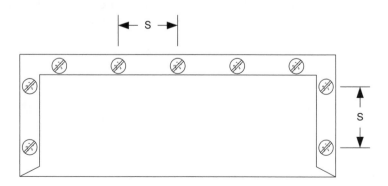

Fig. 5.6 *The distance S between screws should be a small fraction of a wavelength at the lowest frequency protected.*

Fig. 5.7 *"Finger" style box.*

and spurs showing on the spectrum analyzer display began to rise. He eventually reached the screw spacing recommended by the mechanical engineer . . . and at high frequencies the shielding was almost ineffective.

Another form of box is shown in Figure 5.7. This type of box uses a bottom shell that is enclosed on all sides but the top. A top cover with RF "fingers" can be used to shield the top side. The fingers dig into the metal of the bottom shell, creating a tighter RF bond. One popular form of this type of box is manufactured by SESCOM and is made of tinned steel.

Slots

Be wary of slots in shielding enclosures. They are relatively efficient radiators . . . so much so that some microwave antennas are little more than arrays of slot apertures. When the slot approaches 1/8 wavelength or longer, it may radiate rather effectively. This could occur when connectors such as the "DB-x" type used for digital interfaces (e.g., RS-232C) are mounted to the shielded enclosure (Figure 5.8).

Connectors are not the only form of "slot" found in some equipment. If covers or shell halves in aluminum project boxes are just butted together, as shown previously in Figure 5.4, then the lack of a tight fit might form a radiating slot. The best solution is to use boxes with an overlapping "lip" to join the halves together (Figure 5.5). Other accidental slots are created when internal shielding panels are put in place to create multiple shielded compartments, and the mechanical

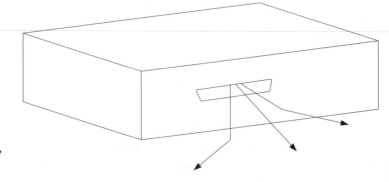

Fig. 5.8
Connectors and other holes in the box are sources of EMI.

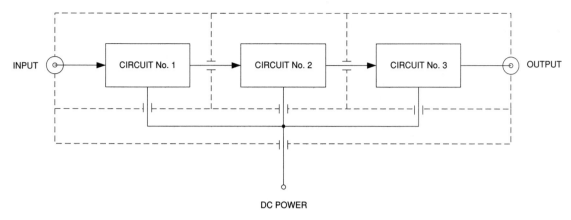

Fig. 5.9 *Multicircuit shielding.*

fit is not good. One reason to use copper or brass to make enclosures is that a bead of solder can be used to ensure that these panels are firmly anchored to ground with no "slotting" effects.

DOUBLE SHIELDING

If you delve into very sensitive equipment, such as receivers and scientific instruments, you will find certain critical circuits double-shielded. The reason is that each shield will produce a reduction of signal by 60 to 100 dB (although the latter requires very good shielding). Let's assume that the run-of-the-mill shield will provide 60 dB of attenuation. If two such shields are provided, one inside the other, then the total attenuation will be on the order of 120 dB. This is the reason why very sensitive or very high-gain instruments use double shielding, especially in their front-end circuitry.

MULTICOMPARTMENT SHIELDING

In some sensitive instruments it is necessary to provide multiple internal shielded compartments to isolate circuits from each other and from the external environment. In Figure 5.9 three circuits are placed inside individual shielded compartments within an overall shielded box. Shielding Rule No. 2 is:

RULE OF THUMB: The number of shielded compartments required is equal to the number of individual circuits that must be protected, plus one for each power entrance.

SPRAY-ON SHIELDING

A lot of equipment today is built in plastic or other synthetic nonconducting forms of cabinet. Unfortunately, these cabinets are an EMI nightmare. In some cases, the manufacturer may apply a conductive coating to the inside of the plastic case to provide shielding. Copper, aluminum, and silver conductive sprays and paints are known. However, they don't always provide a very good shield, so care must be taken. First, of course, is to make sure that the material selected is intended for making shielding. Not all silver, copper, or aluminum paints are truly conductive. And many such paints are not intended for shielding, so they may produce a metal density and thickness that are insufficient. The best one can say about some products is that they are "better than nothing" . . . but not much.

CONNECTORS, METERS, AND DIALS

Objects that penetrate a front or rear panel, or go through the wall of a shielded compart-

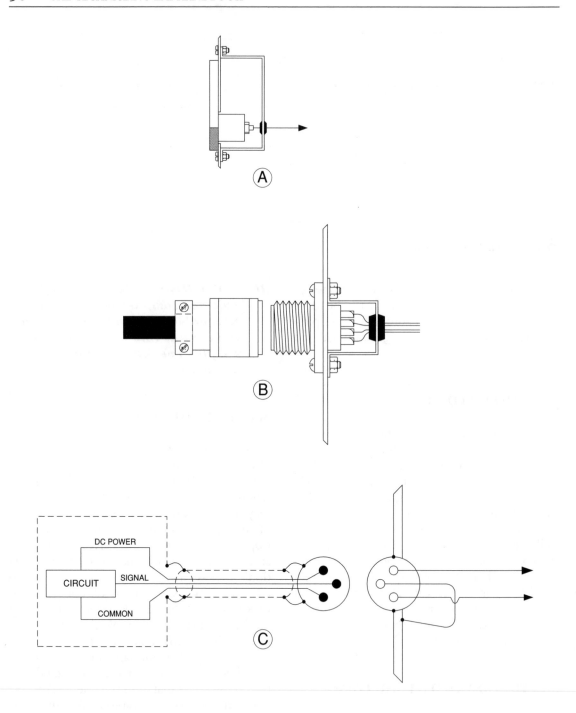

Fig. 5.10 *(A) Meter protection; (B) connector protection; (C) shielded connector.*

ment, pose special problems of shielding. Figure 5.10 shows several different problems and how they can be handled. The case of a digital or analog meter is shown in Figure 5.10A. The large hole cut in the panel for the meter can provide a means for EMI to enter or leave the compartment. A shielding cover is placed over the rear of the meter movement, completely enclosing it except for a small hole with grommet that allows the DC wires to pass in and out. If only DC is carried by the wire, then it's also possible to use feedthrough capacitors to carry the wires (see Chapter 6).

In some other instances a connector for multiconductor cable is used (Figure 5.10B). The cable selected should be a shielded multiconductor type. The rear of the connector is treated in pretty much the same way as the meter movement. For many forms of connector the manufacturer will offer optional EMI shields, but for others a shield will have to be formed.

Figure 5.10C shows the wiring for a multiconductor shielded cable and connector. Assume that a generic circuit or signal source is inside a shielded housing, and it is connected to a shielded instrument through the cable. The signal, common, and DC power supply lines are fed through separate conductors, while the shield for the source end is connected to the shield of the cable. At the end where the connector on the cable mates with the connector on the equipment panel, the same thing is repeated: shield of cable to shield of connector, with the wires kept separate. At some point just inside the cabinet it may be that the ground and shield are connected together with the cabinet shield.

INSTALLING A COAXIAL CONNECTOR

I've messed up my share of RG-8/U and RG-11/U coaxial cable in my time. Putting a PL-259 "UHF" coaxial connector on the end of a length of coax appears to be a daunting task. But it's really quite easy. Figure 5.11 shows the steps. First, cut the cable to the length needed. In Figure 5.11A we see the first connector step: slitting the outer insulation. Take care to slit *only* the outer insulation and not damage the shield braid beneath it. Use a razor knife, hobby knife, or scalpel to make a 1-1/8 inch (30mm) slit from one end of the cable. Next, as shown in Figure 5.11B, make a circumferential cut at the inner end of the long slit, so that the outer insulation can be removed (Figure 5.11C).

Once the outer insulation is removed, use a soldering iron to lightly tin the braid, making it stiff. Take care to not use too much heat, or the inner insulation will be damaged. Also, if too thick a layer of solder is laid on, then the connector will not slip over the end in the following steps.

Figure 5.11D shows the next step: cut away the tinned outer braid and inner insulation so that 0.75 inch (19 mm) of inner conductor is exposed. The inner insulator and shield will be flush with each other at the end of the 19-mm section.

Figure 5.11E shows the PL-259 UHF connector and the prepared cable. Now here's a bit of wisdom: unless the cable is relatively short, and you have free access to both ends, now is the time to slip the outer shell (the piece with the thumb threads) over the cable, back first. Next, slip the inner portion of the connector over the exposed cable end, making sure that the inner conductor goes into the hollow connector center pin, without bending or buckling (Figure 5.11F). Some people prefer to solder tin the inner conductor, especially if the cable uses a stranded wire for the inner conductor.

Once the connector is firmly seated on the cable (Figure 5.11G), use an ohmmeter to make sure there is no short circuit between inner and outer conductors. If the cable checks out on the ohmmeter, solder the outer shield to the connector by soldering through the view holes in the thinner portion of the connector. Make sure that *all* of the holes are soldered. I've seen a connector "tacked on" by soldering only one hole go bad. The connection failed . . . and the blankety-blank cable was buried inside a wall (which itself was an installation mistake!). Finally, solder the inner conductor at the end of the hollow center pin. After it is cooled, check the cable with an ohmmeter.

I use a soldering gun with 100/200/250-watt three-way switchable settings, rather than a pencil iron. The pencil iron is fine for printed circuit boards, but the metal of the braid and connector heat-sink enough to cool off the connection too much. Give it a try . . . the connector is actually relatively easy to install.

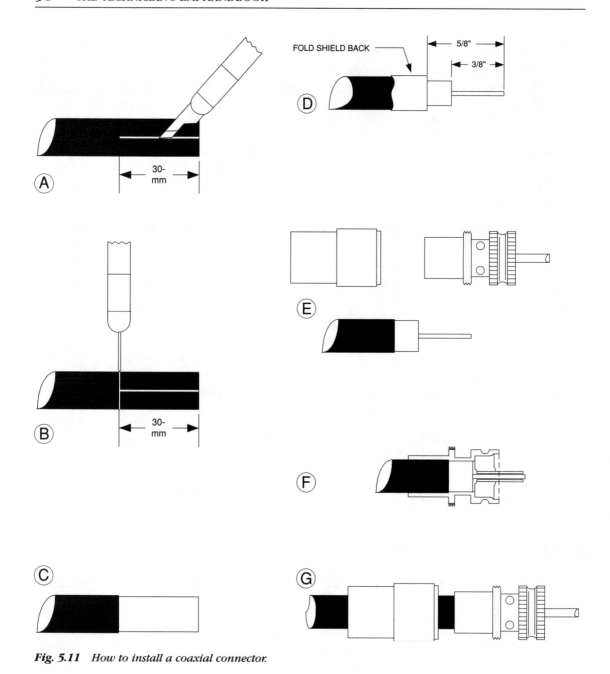

Fig. 5.11 *How to install a coaxial connector.*

GUARD SHIELDING

One of the properties of the differential amplifier, including the instrumentation amplifier, is that it tends to suppress interfering signals from the environment. The common-mode rejection process is at the root of this capability. When an amplifier is used in a situation where it is connected to an external signal source through wires, those wires are subjected to strong local 60-Hz AC fields from nearby power-line wiring. Fortunately, in the case of the differential amplifier the field affects both input lines equally, so the induced interfering signal is cancelled out by the common-mode rejection property of the amplifier.

Unfortunately, the cancellation of interfering signals is not total. There may be, for example, imbalances in the circuit that tend to deteriorate the CMRR of the amplifier. These imbalances may be either internal or external to the amplifier circuit. Figure 5.12A shows a common scenario. In this figure we see the differential amplifier connected to shielded leads from the signal source, V_{in}. Shielded lead wires offer some protection from local fields, but there is a problem with the standard wisdom regarding shields: it is possible for shielded cables to manufacture a

valid differential signal voltage from a common-mode signal!

Figure 5.12B shows an equivalent circuit that demonstrates how a shielded cable pair can create a differential signal from a common-mode signal. The cable has capacitance between the center conductor and the shield conductor surrounding it. Input connectors and the amplifier equipment internal wiring also exhibit capacitance.

These capacitances are lumped together in the model of Figure 5.12B as C_{S1} and C_{S2}. As long as the source resistances and shunt resistances are equal, and the two capacitances are equal, there is no problem with circuit balance. But inequalities in any

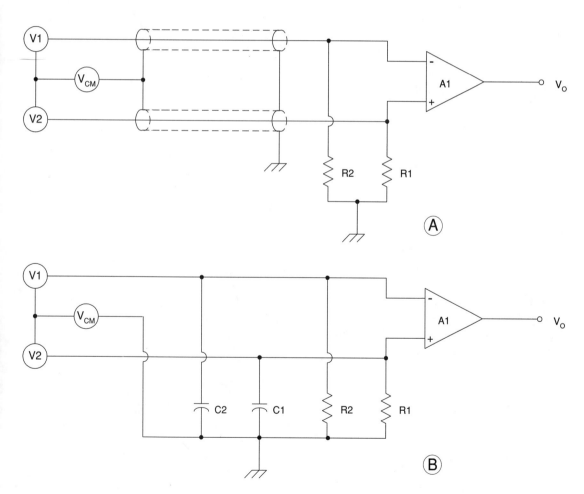

Fig. 5.12 *(A) Differential input circuit; (B) equivalent circuit.*

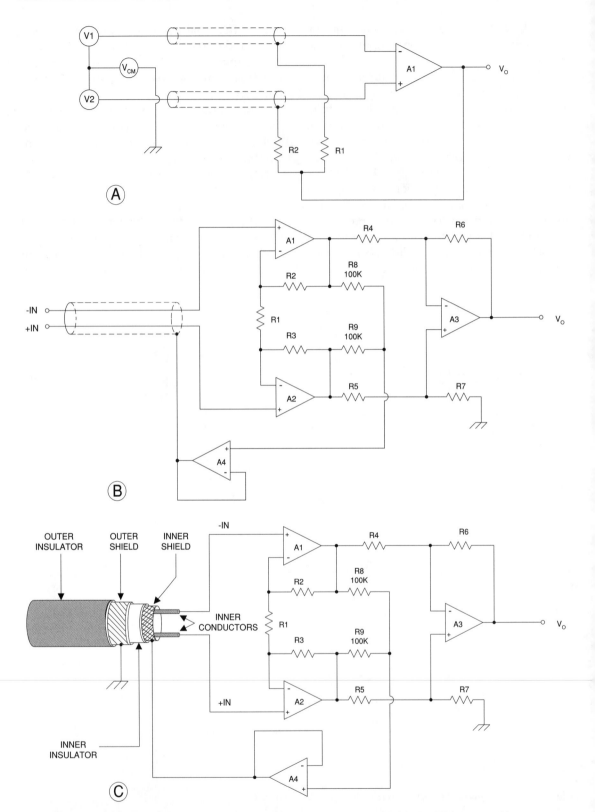

Fig. 5.13   *(A) Common-mode protection; (B) better circuit; (C) use of a double-shielded input cable.*

of these factors (which are commonplace) creates an unbalanced circuit in which common-mode signal V_{cm} can charge one capacitance more than the other. As a result, the difference between the capacitance voltages, V_{CS1} and V_{CS2}, is seen as a valid differential signal.

A low-cost solution to the problem of shield-induced artifact signals is shown in Figure 5.13A. In this circuit a sample of the two input signals is fed back to the shield, which in this situation is not grounded. Alternatively, the amplifier output signal is used to drive the shield. This type of shield is called a guard shield. Either double shields (one on each input line) as shown or a common shield for the two inputs can be used.

An improved guard shield example for the instrumentation amplifier is shown in Figure 5.13B. In this case a single shield covers both input lines, but it is possible to use separate shields. In this circuit a sample of the two input signals is taken from the junction of resistors R8 and R9 and fed to the input of a unity gain buffer/driver "guard amplifier" (A4). The output of A4 is used to drive the guard shield.

Perhaps the most common approach to guard shielding is the arrangement shown in Figure 5.13C. Here we see two shields used; the input cabling is double-shielded insulated wire. The guard amplifier drives the inner shield, which serves as the guard shield for the system. The outer shield is grounded at the input end in the normal manner and

serves as an electromagnetic interference suppression shield.

Guard Shielding on Printed Circuit Boards

The guard shield concept can be extended to the input pins of high-gain amplifiers on printed circuit boards. Figure 5.14 shows the method for placing a guard ring around a printed circuit pad (e.g., an IC pin). The ring is grounded in most cases, but is always connected to the guard shield of the input cable. The center conductor of the cable is connected to the pad itself. The version shown in Figure 5.14 keeps one end of the ring open in order to accommodate connection to the PC pad. If the board is two-sided or multilayered, then the ring can be complete, with the connection to the pad occurring on the top or on one of the intermediate layers.

Guard Shielding on Specialized Amplifiers

There are a number of specialist integrated circuits on the market. These are often designed specially for the analog subsystem of PC-based instruments, although in many cases they are more general in format. An example of a general IC amplifier is the B-B INA-101AG device shown in Figure 5.15. It is an integrated circuit instrumentation amplifier (ICIA) device, with the gain being set by an external resistor R_G. This type of device has provision for guard shield connections directly to the IC (pins 4 and 8). These pins are fed to the separate in-

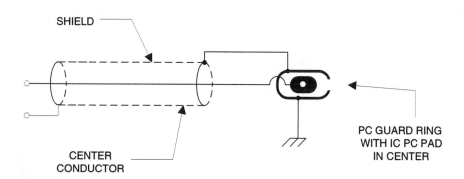

Fig. 5.14 *Shielded PCB eyelet.*

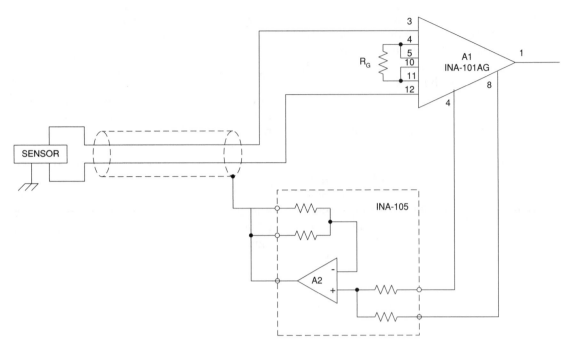

Fig. 5.15   *Sensor input circuit.*

puts of a summing amplifier, the output of which drives the guard shield.

GROUNDING AND GROUND LOOPS

Impulse noise due to electrical arcs, lightning bolts, electrical motors, and other devices can interfere with the operation of sensors and their associated circuits. Shielding of lines (see earlier discussion) will help somewhat, but it isn't the entire answer. Filtering (discussed next) is useful, but it is at best a two-edged sword, and it must be done prudently and properly. Filtering on signal lines tends to broaden fast-rise-time pulses and attenuate high-frequency signals . . . and in some circuits causes as many problems as it solves.

Other electrical devices nearby can induce signals into the instrumentation system, the chief among these sources being the 60-Hz AC power system. It is wise to use only differential amplifier inputs, because of their high common-mode rejection ratio. Signals from the desired source can be connected across the two inputs, and so become a differential signal, while the 60-Hz AC interference tends to affect both inputs equally (and so is common mode).

It is sometimes possible, however, to manufacture a differential signal from a common-mode signal. Earlier we said this phenomenon was due to bad shielding practices. In this section we are going to expand on that theme and consider grounds as well as shields.

One source of this problem is called a ground loop and is shown in Figure 5.16A. This problem arises from the use of too many grounds. In this example the shielded source, shielded input lines, amplifier, and DC power supply are all grounded to different points on the ground plane. Power-supply DC (I) flow from the power supply at point A to the amplifier power common at point E. Since the ground has ohmic resistance, albeit very low resistance, the voltage drops E_1 through E_4 are formed. These voltages are seen by the amplifier as valid signals and can become especially troublesome if I is a varying current.

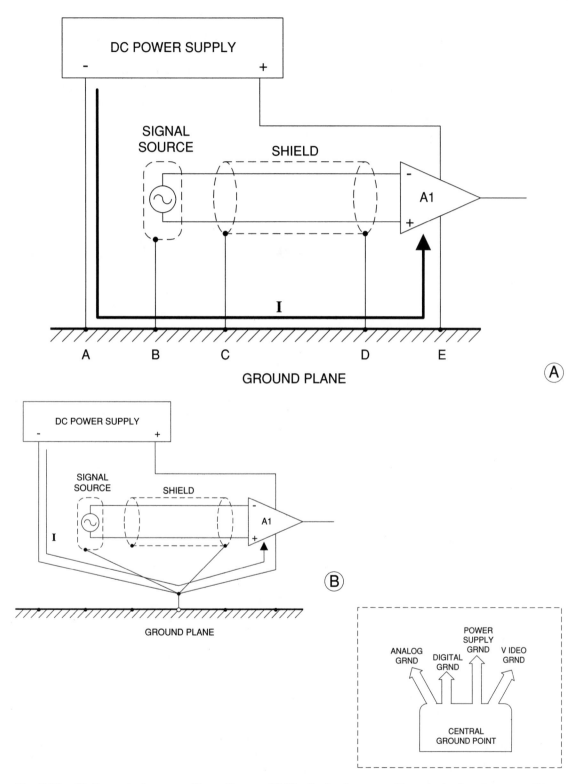

Fig. 5.16   *(A) Scenario for ground-loop disaster. (B) The fix for the ground loop is single-point grounding at low frequencies.*

The solution to this problem is to use single-point grounding, as shown in Figure 5.16B. Some amplifiers used in sensitive graphics or CRT oscilloscopes keep the power and signal grounds separate, except at some single, specific common point. In fact, a few models go even further by creating several signal grounds, especially where both analog and digital signals might be present.

In some instances the shield on the input lines must not be grounded at both ends. In those cases, it is usually better to ground only the amplifier ends of the cables.

Chapter 6

Filtering Electronic Circuits

In Chapter 5 we looked at shielding the circuit in order to protect it from EMI/RFI. Now let's look at the rest of the story: filtering. Electronic circuits must perform two different functions: (1) they must respond to desired signals, and (2) they must reject undesired signals. But the world is full of a variety of interfering signals, all of which are grouped under the headings electromagnetic interference (EMI) or radio-frequency interference (RFI). These EMI/RFI sources can ruin the performance of, or destroy, otherwise well-functioning electronic circuits. Oddly, most circuits are designed for function 1: they do, in fact, respond properly to desired signals. But many, many devices fail miserably on function 2: they will, in fact, respond to undesired signals in an inappropriate manner. Let's take a look at some of the techniques that can be used to EMI/RFI-proof electronic devices and circuitry.

SHIELDING

The first step in providing protection against EMI/RFI is shielding the circuitry. The entire circuit is placed inside a metallic enclosure that prevents external EMI/RFI fields from in-teracting with the internal circuits, and also prevents internal fields from doing the opposite. Let's assume for this discussion that the circuits are well shielded (see Chapter 5). So what's the problem? Unfortunately, the shield is not perfect because of power and signal leads entering and/or leaving the enclosure. For those, some sort of filtering is needed.

BASIC TYPES OF FILTERS

Filters come in a variety of types, but perhaps the most useful way of categorizing them is by passband: low-pass, high-pass, bandpass, and notch filters are the basic classes.

Low-Pass Filters (LPFs)

The low-pass frequency response (Figure 6.1A) passes all frequencies below a critical cutoff frequency (F_c) and attenuates those above F_c. The cutoff frequency is usually defined as the point at which the gain falls off -3 dB from the midband response or, if the response is uneven, the response at a defined frequency. The cutoff is not abrupt, but rather will fall off at a given slope. This roll-off

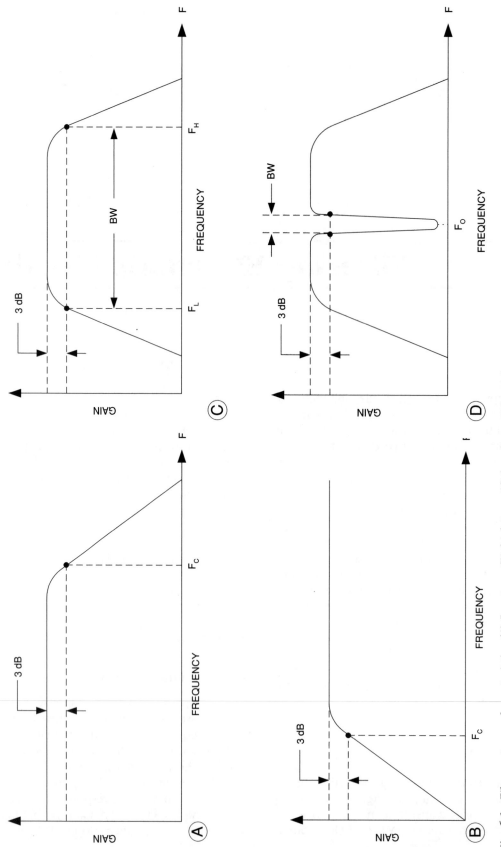

Fig. 6.1 Filter response characteristics: (A) Low-pass; (B) high-pass; (C) bandpass; (D) notch.

66

above F_c is usually defined in terms of decibels per decade (a 10:1 frequency change) or decibels per octave (a 2:1 frequency change). The low-pass filter is used to attenuate EMI/RFI signals above F_c.

High-Pass Filters (HPFs)

The high-pass response (Figure 6.1B) is exactly the opposite of the low-pass response. The filter passes frequencies above F_c, but attenuates those frequencies below F_c. Again, there is a roll-off slope below F_c.

Bandpass Filters (BPFs)

The bandpass filter response (Figure 6.1C) is essentially overlaid LPF and HPF responses. There are two cutoff frequencies: a lower limit (F_L) and an upper limit (F_H), both of which are defined at −3 dB points. The bandwidth of the BPF is defined as the difference between −3 dB points:

$$\text{B.W.} = F_H - F_L \qquad (6.1)$$

In most cases, the BPF will have a center frequency (F_O) specified. The Q or "quality factor" of the BPF is defined as the ratio of the center frequency to bandwidth, or:

$$Q = \frac{F_O}{\text{B.W.}} \qquad (6.2)$$

where Q is dimensionless, and the two other terms are expressed in the same units.

Notch Filters

The notch filter response (Figure 6.1D) is one that has a very high attenuation at a specific frequency within the circuit's passband, but passes the other frequencies. In Figure 6.1D the notch is superimposed over a BPF response, although it may also be found with LPF, HPF, or wideband (i.e., flat) responses. The notch filter is used to take out a specific interfering frequency. For example, if 60-Hz AC power-line interference is terribly bothersome, then a notch filter might be used.

FILTER CIRCUITS

Filter circuits can be active or passive, but for this present discussion let's consider the passive varieties only. Such filters are made of either R-C, L-C, or R-L-C components in appropriate types of network.

Figure 6.2 shows simple resistor-capacitor (R-C) networks in both low-pass and high-pass filter configurations. Note that the two circuits are similar except that the positioning of the R and C components are re-

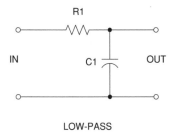

LOW-PASS

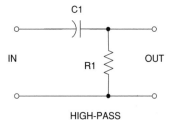

HIGH-PASS

$$F_C = \frac{1}{2\pi\ R1\ C1}$$

Fig. 6.2 *Resistor–capacitor (R-C) low-pass and high-pass filter circuits.*

versed. The cutoff frequency of these circuits is found from:

$$F_C = \frac{1}{2\,\pi\,R1\,C1} \qquad (6.3)$$

where F_C is the cutoff frequency in Hertz (Hz), $R1$ is in ohms, and $C1$ is in farads. These circuits provide a frequency roll-off beyond F_C of –6 dB/octave, although sharper roll-off can be obtained by cascading two or more sections of the same circuit.

Figure 6.3 shows four different Chebychev filters (two LPF and two HPF).

Each of these filter circuits is a "five-element" circuit, i.e., each has five L or C components. Fewer (e.g., three) and more (e.g., seven and nine) elements are also used. Fewer elements will give a poorer frequency roll-off, while more elements give a sharper roll-off. The component values given in these circuits are normalized for a 1-MHz cutoff frequency. To find the required values for any other frequency, divide these values

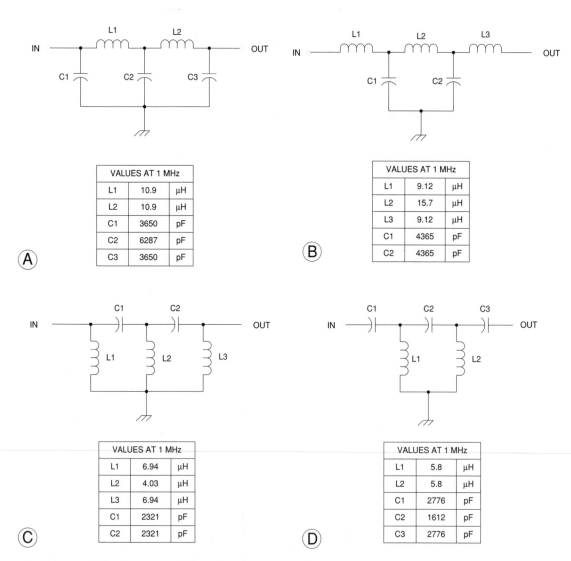

Fig. 6.3 *L-C filter circuits: (A) LPF pi-configuration; (B) LPF tee-configuration; (C) HPF pi-configuration; (D) HPF tee-configuration.*

by the desired frequency in megahertz. For example, to make a high-pass tee-configuration filter for, say, 4.5 MHz, take the values of Figure 6.3D and divide by 4.5 MHz:

$$L1 = 5.8 \text{ µH}/4.5 \text{ MHz} = 1.29 \text{ µH}$$
$$L2 = L1$$
$$C1 = 2776 \text{ pF}/4.5 \text{ MHz} = 617 \text{ pF}$$
$$C2 = 1612 \text{ pF}/4.5 \text{ MHz} = 358 \text{ pF}$$
$$C3 = C1$$

If the desired frequency is less than 1 MHz, then it must still be expressed in megahertz, i.e., 100 kHz = 0.1 MHz and 10 kHz = 0.01 MHz.

R-C EMI/RFI PROTECTION

Some circuits, especially those that operate at low frequencies, may use R-C low-pass filtering for the EMI/RFI protection function. Consider the differential amplifier in Figure 6.4. This circuit is representative of a number of scientific and medical instrument amplifier input networks. A medical electrocardiogram (ECG) amplifier, for example, is basically a differential amplifier with a high gain (1,000 to 2,000) and a low frequency response (0.05 to 100 Hz). It picks up the human heart's electrical activity as seen from skin electrodes on the surface.

There are a number of problems that will afflict the recording, other than the obvious 60 Hz problem. The ECG must often be used in the presence of strong radio-frequency (RF) fields from electrosurgery machines. These "electronic scalpels" are used by surgeons to cut and cauterize and will produce very strong fields on frequencies of 500 kHz to 3 MHz. They must also survive high-voltage DC jolts from a charge stored in a capacitor when the patient must be resuscitated. The defibrillator machine is used to "jump start" a patient's heart that is in ventricular fibrillation (a fatal arrhythmia). It will produce short-duration voltage spikes ranging from hundreds of volts to several kilovolts, depending on the particular waveform design and energy setting. And those potentials might be applied directly across the ECG amplifier, placing it at risk.

Figure 6.4 contains both RF filtering and a means for limiting the defibrillator jolt. The resistors and capacitors form a three-stage cascade RC filter, one for each input of the differential amplifier. These components will filter the RF component. Typical values range from 100K to 1 megohm for the resistors, and 100 pF to 0.01 µF for the capacitors. The values should not produce a cutoff frequency of 100 Hz or less. The high-voltage protection is provided by a combination of the input resistors and a pair of zener diodes (D1 and D2) shunting the signal and common lines. In some older ECG amplifiers, NE-2 neon glow-lamps were used in place of the zener diodes.

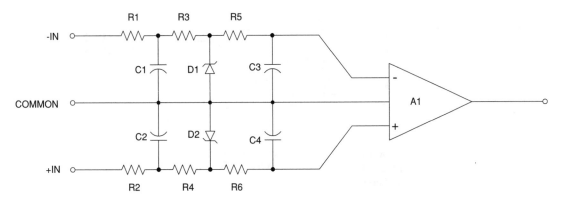

Fig. 6.4 *R-C and high-voltage protection of low-frequency circuits such as ECG amplifiers.*

FEEDTHROUGH CAPACITORS

One effective way to reduce the effects of EMI/RFI that pass into a shielded compartment via power and signal lines is the feedthrough capacitor (Figure 6.5). These ca-

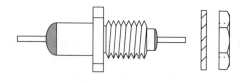

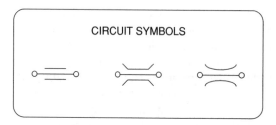

CIRCUIT SYMBOLS

Fig. 6.5   *Feedthrough capacitor.*

pacitors typically come in 500-, 1,000-, and 2,000-pF values. Both solder-on and screw-on (shown in Figure 6.5) types are available. In some catalogs these capacitors are referred to as "EMI filters" rather than "feedthrough capacitors." I may be a cynic, but the "EMI" designation seems to add considerably to the price without any apparent advantage over straight feedthrough capacitors. Also shown in Figure 6.5 are several different forms of circuit symbols used for feedthrough capacitors in circuit diagrams.

There are several different ways to use a feedthrough capacitor. One method is to simply pass it through the shielded compartment wall and attach the wires to each side. In other cases, additional resistors or inductors are used to form a low-pass filter. Figure 6.6 shows one approach in which a radio-frequency choke (RFC) is mounted external to the shielded compartment. This method is often used for TV and cable-box tuners. One end of the RFC is connected to the feedthrough capacitor, and the other to some other point in the external circuit (e.g., a stand-off insulator is shown here, as is common on TV/cable tuners, but other points are used as well). It is very im-

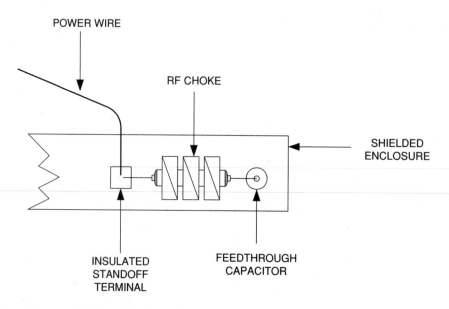

POWER WIRE

RF CHOKE

SHIELDED ENCLOSURE

INSULATED STANDOFF TERMINAL

FEEDTHROUGH CAPACITOR

Fig. 6.6   *Use of external RFC and feedthrough capacitor.*

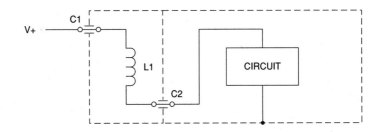

Fig. 6.7
Built-in filter.

portant to keep the lead wire from the RFC to the feedthrough capacitor as short as possible to limit additional pickup beyond the filtering.

Another approach is shown in Figure 6.7. This method uses a separate shielded compartment inside the main shielded enclosure. Feedthrough capacitors ($C1$ and $C2$) are used to carry DC (or low-frequency) signals into and out of the filter compartment. An inductor, $L1$, is part of the filtering, so the combination $L1$–$C1$–$C2$ forms a low-pass pi-configuration filter. The inductor may also be an RF choke, but the effect is the same.

One caution is in order: if L-C filters are used on both input and output signal lines of a circuit, then make sure they resonate on different frequencies. The reason is that they will form a tuned-input/tuned-output oscilla-tor if the circuit being protected has sufficient gain at the filter's resonant frequency.

In some cases, connectors are bought that have filtering built-in (Figure 6.8). These products are usually described as EMI filtering connectors. Although most of them are designed to work with 120/220V AC power lines, others are available that work at higher frequencies.

One more approach is shown in Figure 6.9. This application is a little harder to see because the "filtering" is performed by using a set of one or more ferrite beads slipped over the wire from the connector pin to the circuit board. Ferrite beads surrounding a wire act like a small-value RF choke, and so will filter (typically) VHF/UHF frequencies. It is common to see these beads on RF equipment, but they are also found on digital devices.

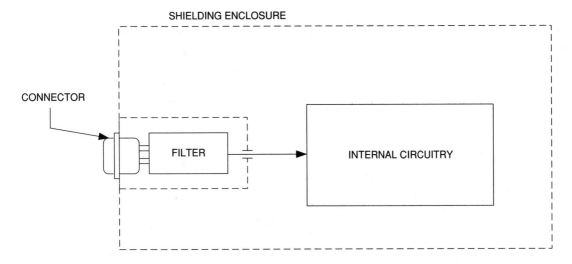

Fig. 6.8 ***Connector/EMI-filter combination.***

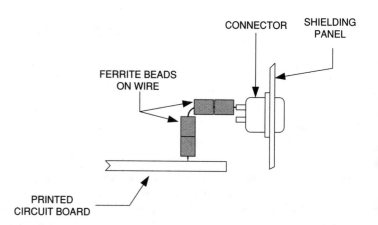

Fig. 6.9
Use of ferrite beads as RF choke to protect pins from a connector.

NOTCH FILTERING

One of the more difficult interference problems is in-band interference. In radio receivers the IF passband may not be totally effective in eliminating unwanted near-channel or cochannel signals. Alternatively, an undesired station may create a heterodyne beat with the desired signal. These spurious signals are generated in the receiver and produce an audio tone equal to the difference in frequency between the two RF signals.

In medical and scientific instruments, radiation from the 60-Hz AC power mains gets into circuits and causes problems. A medical electrocardiogram (ECG) recording has a peak amplitude on the order of 1 mV and requires a bandpass of 0.05 to 100 Hz for proper recording of the waveform. Guess where the AC power-line interface is found . . . right in the middle of the band.

The usual solution to unwanted in-band interfering frequencies is the notch filter. The frequency response of a typical notch filter is shown in Figure 6.10 in greater detail than given earlier. These filters are similar to another class, i.e., bandstop filters, but the band of rejection is very narrow around the center frequency (F_c). The bandwidth (BW) of these filters is the difference between the frequen-

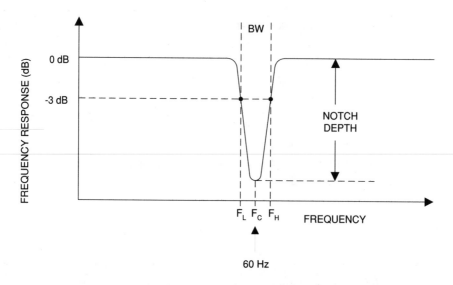

Fig. 6.10 *Notch filter response in better detail.*

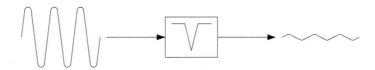

Fig. 6.11 *Effect of notch filter.*

cies at the two –6 dB points, when the out-of-notch response is the reference 0 dB point. These frequencies are F_L and F_H, so the bandwidth is $F_H − F_L$.

The "sharpness" of the notch filter is a measure of the narrowness of the bandwidth and is specified by the "Q" of the filter. The Q is defined as the ratio of the center frequency (F_c) to bandwidth (BW), just as it is for bandpass filters.

For example, a notch filter that is centered on 50 Hz and has –6 dB points at 48 and 52 Hz (4-Hz bandwidth) has a Q of 50/4 or 12.5.

The notch filter does not remove the entire offending signal, but rather suppresses it by a large amount (Figure 6.11). The notch depth (see again Figure 6.10) defines the degree of suppression and is defined by the ratio of the gain of the circuit at an out-of-notch frequency (e.g., F_{ob}) to the gain at the notch frequency. Notch depths of –30 to –50 are relatively easily obtained, and deeper notches (e.g., –60 dB) are possible. Assuming equal input signal levels at both frequencies (which has to be checked; most signal generators have variable output levels with changes of frequency!), the notch depth can be calculated from the output voltages of the filter at the two different frequencies:

$$\text{Notch depth} = 20 \log \left(\frac{V_{FC}}{V_{OB}} \right) \quad (6.4)$$

The depth can also be calculated by applying the same equation using the filter input and output voltages at the notch frequency.

TWIN-TEE NOTCH FILTER NETWORKS

One of the most popular forms of notch filter is the twin-tee filter network, shown in Figure 6.12. It consists of two R-C T-networks, consisting of $C1/C3/R2$ and $R1/R3/C2$. Notch depths of –30 to –50 dB are easy to obtain with the twin-tee, assuming proper circuit design and component selection. Very good matching and selection of parts makes it possible to achieve –60 dB suppression.

The center notch frequency of the network in the generic case is given by:

$$F_c = \frac{1}{2\pi} \sqrt{\frac{C1 + C3}{C1\ C2\ C3\ R1\ R3}} \quad (6.5)$$

We can simplify this expression by adopting a convention that calls for the following relationships:

$C1 = C3 = C$
$R1 = R3 = R$
$C2 = 2C$
$R2 = R/2$

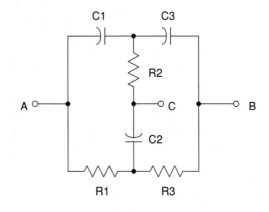

SIMPLIFIED EXPRESSION

C1 = C3 = C
C2 = C
R1 = R3 = R
R2 = R/2
FC = 1/6.28RC

Fig. 6.12 *Twin-tee network.*

If this convention is adopted, then we can reduce the frequency equation to:

$$F_c = \frac{1}{2\pi RC} \qquad (6.6)$$

In these expressions, F is in hertz (Hz), R is in ohms, and C is in farads. Be sure to use the right units when working these problems: "10 kohms" is 10,000 ohms, and "0.001 μF" is 1×10^{-9} farads. In calculating values, it is usually prudent to select a capacitor value, and then calculate the resistance needed. This is done for two reasons: first, there are many more standard resistance values, and second, potentiometers can be easily used to trim the values of resistances, but it is more difficult to use trimmer capacitors.

One of the problems of these filters is that the depth of the notch is a function of two factors involving these components: first, that they be very close to the calculated values, and second, that they be matched closely together. The capacitors can be selected at random from a group of a dozen or so "mine run" capacitors of good quality and matched to the required value using a digital capacitance meter such as those found on digital multimeters. Real odd values can be obtained by parallel or series connection of two or more capacitors. The resistors can be selected 5 percent metal film resistors or 1 percent precision resistors.

The order of priority of selection is to find those that closely matched each other, and only incidentally how close they come to the calculated value. Errors in the mean capacitance of the selected group can be trimmed out using a potentiometer in the resistor elements of the twin-tee network.

When selecting a frequency source, either select a well-calibrated source, or use a frequency counter to measure the frequency. Keep in mind the situation described earlier where only a 2-Hz shift produced a 38-dB difference in notch depth! Alternatively, use a 6.3V or 12.6V AC filament transformer secondary as the signal source.

WARNING

The primary circuits of these transformers are at a potential of the AC line, and can thus be lethal if mishandled!

Adjustable frequency notch filters can be built using the twin-tee idea, but none of the usual solutions are really acceptable. One implementation requires three ganged matched potentiometers or three ganged capacitors. Unfortunately, in either case at least one of the variable sections must be of different value from the other two, causing a tracking problem that spreads the notch. You might not notice a tracking problem in some circuits, but in a high-Q notch filter it can be disastrous.

ACTIVE TWIN-TEE NOTCH FILTERS

Active frequency selective filters use an active device such as an operational amplifier to implement the filter. In the active filter circuits to follow, the "twin-tee" networks are shown as block diagrams for the sake of simplicity and are identical to those circuits shown earlier; the ports "A," "B," and "C" in the following circuit are the same as in the previous network.

The simplest case of a twin-tee filter is to simply use it "as is," i.e., use the filter circuits shown earlier. But the better solution is to include the twin-tee filter in conjunction with one or more operational amplifiers. There is one solution in which the twin-tee network is cascaded with an input buffer amplifier (optional) and an output buffer amplifier (required). These amplifiers tend to be noninverting op-amp follower circuits. The purpose of buffer amplifiers is to isolate the network from the outside world. For low-frequency applications, the op-amps can be 741, 1458 and other similar devices. For higher frequency applications, i.e., those with an upper cutoff frequency above 3 kHz, use a non-frequency-compensated device such as the CA-3130 or CA-3140 devices.

A superior circuit is shown in Figure 6.13. In this circuit, port C of the twin-tee network (the common point) is connected to the output terminal of the output buffer amplifier. There is also a feedback network consisting of two resistors (R_a) and a capacitor (C_a). The values of R and C in the twin-tee network are found from the equation above, while the values of R_a and C_a are found from:

$$R_a = 2RQ \qquad (6.7)$$

and

$$C_a = \frac{C}{Q} \qquad (6.8)$$

A variable-Q control notch filter is shown in Figure 6.14. In this circuit, a noninverting follower (A3) is connected in the feedback loop in place of R_a and C_a. The Q of the notch is set by the position of the 10 kohm potentiometer ($R2$). Values of Q from 1 to 50 are available from this circuit.

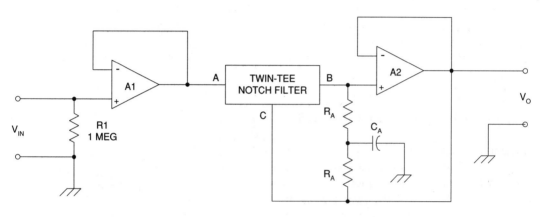

Fig. 6.13 *Active twin-tee notch filter circuit.*

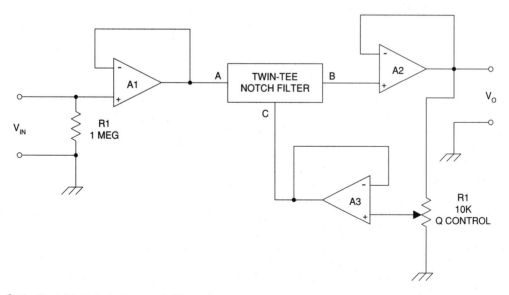

Fig. 6.14 *Variable-Q twin-tee notch filter.*

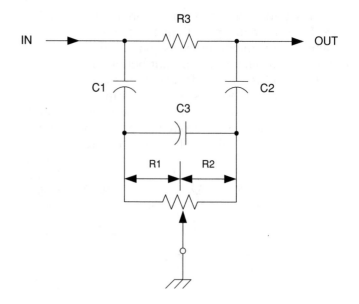

Fig. 6.15
Bridged-tee network.

ADJUSTABLE BRIDGED-TEE CIRCUITS

A variant of the bridged-tee notch filter is shown in Figure 6.15. This circuit is often used in cases where the notch frequency is either variable or not known with great precision. One popular use for this filter is on radio receivers, where it is used to notch unwanted audio tones in the output. The notch frequency is given by:

$$F = \frac{1}{2\pi C \sqrt{3\, R1\, R2}} \qquad (6.9)$$

assuming:

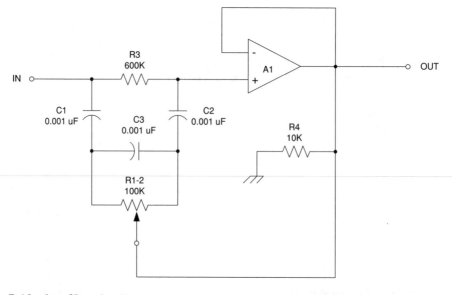

Fig. 6.16   *Bridged-tee filter circuit.*

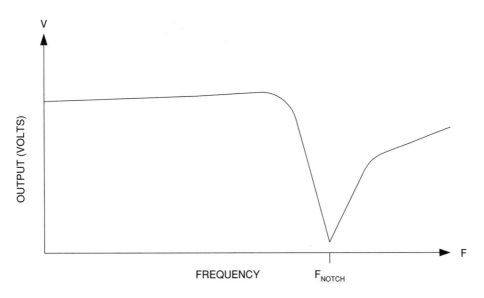

Fig. 6.17 *Frequency response of Figure 6.16.*

$$C1 = C2 = C3 = C$$

R1 and R2 are the wiper-to-end-terminal resistances of the potentiometer where:

F is in hertz (Hz)
C is in farads
R1 and R2 are in ohms

A sample circuit based on Figure 6.15 is shown in Figure 6.16. This circuit will produce a notch between about 1.8 and 8 kHz, depending on the setting of R1. A representative frequency response for this circuit is shown in Figure 6.17.

GYRATOR CIRCUITS

Another approach to notch filter circuits is shown in Figure 6.18. This circuit is sometimes called the gyrator or active inductor notch filter (it's also sometimes called the virtual inductor notch filter).

The notch frequency is set by:

$$F_c = \frac{1}{2\pi \sqrt{R_a \, R_b \, C_a \, C_b}} \qquad (6.10)$$

Equation 6.10 can be simplified to

$$F_c = \frac{1}{2\pi R \sqrt{C_a \, C_b}} \qquad (6.11)$$

if the following conditions are met:

$$\frac{R3}{R1} = \frac{R2}{R_a + R_b} = \frac{R2}{2R} \qquad (6.12)$$

It is possible to use any one of the elements, C_a, C_b, R_a, or R_b, to tune the filter. In most cases, C_a is made variable and C_b is a large value fixed capacitor. The 1,500-pF variable capacitor can be made by paralleling all sections of a three-section broadcast variable, with a single small fixed or trimmer capacitor. Alternatively, since most applications will require a trimmer rather than a "broadcast" variable capacitor, it is also possible to parallel one or more small capacitor and a trimmer. For example, a 100-pF trimmer can be parallel connected with a 1,000-pF and a 470-pF to form the 1,500-pF capacitance required. Make sure that you use low-drift, precision capacitors, or you can match them using a digital capacitance meter.

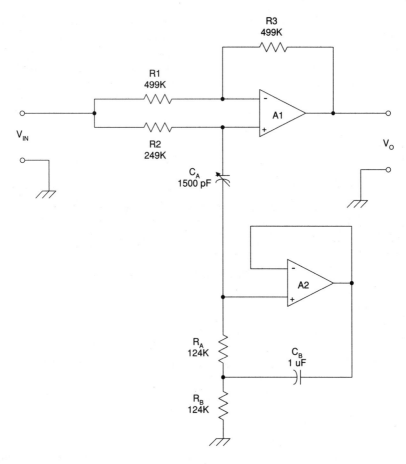

Fig. 6.18   *Gyrator-style notch filter.*

CAUTIONARY NOTE

Be careful when using any filter to remove components from a waveform. If the filter is not a high-Q type, then too much of the signal might be removed. In medical electrocardiograph (ECG) systems the signal has components from 0.05 to 100 Hz, so 60 Hz is right in the center of the range! Oops. To make matters worse, the leads have to be connected to the human body, and so are unshielded at their very ends. Interference from 60 Hz is almost guaranteed unless care is taken. But filtering can take out components that assist the physician in making diagnosis, so it is only used when it is unavoidable. On medical ECG amplifiers the filter is usually switchable so it can be either in or out of the circuit.

GENERAL GUIDELINES

Thus far we've looked at a number of different filtering approaches to protecting equipment. Now let's consider the general guidelines:

1. Always shield the circuit. A circuit that is not shielded cannot be adequately protected by filtering. There is simply too much chance of direct pickup of the EMI/RFI source by the components and wires of the circuit. Also, filtering of EMI/RFI generators (such as transmitters) will not help much if the device is not shielded.

2. Apply filtering to the DC power lines entering or leaving the circuit's shielded enclosure.

3. Use the minimum filtering necessary for accomplishing the level of protection required. It does no good to add one more section of filtering when the job is done properly, but does add cost, complexity, and opportunities for failed components (which will keep the service technician busy, if nothing else).

4. If it is necessary to filter signal input and output lines, use the minimum values of capacitance or inductance consistent with the degree of protection needed. Keep the cutoff frequency well away from the frequencies the circuit normally uses. The filtering will affect those frequencies as well as the undesired frequencies, so it is necessary to select values that minimize the effect on desired frequencies while maximizing the effect on undesired frequencies.

CONCLUSION

Filtering the DC power and signal lines entering or leaving a shielded circuit will go a long way toward eliminating any EMI/RFI problems that are experienced.

Chapter 7

AC Power-Line and Electrical Device EMI

Power lines are supposed to be clean and trim . . . right? Wrong, not by a long shot! Power lines are as "dirty" as you please . . . perhaps dirtier. A number of different conducted sources interfere with equipment through the power lines. There are brownouts (when voltage sags to 95 volts in the United States and Canada) and surges (when the voltage increases to 135 volts). Lightning causes its share of havoc, as does ordinary switching transients. In a medical center, we once flunked the entire freshman class of medical students on their physiology exam. Also failed were all of the basic sciences (i.e., Ph.D.) students and allied health students (e.g., nurse anesthesia, nurse practitioner, physician's assistant). Those people are about "up to here" with stress their first year in medical school, so the profs hoped that the results would not get out until we solved the problem. It turned out to be power line noise.

In those days computing was mainframe computing. The examination was taken on "mark sense" optical scanner paper. You know the stuff . . . "use a No. 2 pencil and completely blacken the desired box." In our case, the optical scanner was connected to a keypunch machine. To younger readers: A keypunch was a noisy, clunky machine that looked like a typewriter on steroids that punched the holes in old-fashioned computer cards. The cards were then taken to the computer center for processing overnight. When the computer printout paper was returned, the grades were recorded (manually!) and the paper posted for all to see.

The problem was solved when one of the engineers I worked with noticed that one column on the computer card had all digits punched out. There is no EBCDIC code that has all digits punched out in a single column. The problem was traced to high-voltage power-line transients arising from load shifting switching gear in the basement. It seems that the local power company gave the university a 2-percent break on the electric bill if they installed equipment that would periodically balance the load between the three phases (which makes for more efficient operation). Unfortunately, the TRIAC switches tossed huge (>2-kV) transients that averaged 50 to 100 microseconds.

The solution to the problem (we couldn't rewire the building or turn off the load switchers) was to place a *Topaz* isolation transformer between the power line and the optical scanner and keypunch machine. These transformers are specially designed to snuff power-line noise (today, we might use a computer surge suppressor for many such applications).

We then found that the noise was the basis for a lot of problems. For example, the electron microscope guy had been attempting to find a "vibration problem" in his equipment (it didn't help that the subway ran right beneath our building foundation . . . so he was tuned in on "vibration problems"). Adding an isolation transformer to his equipment cured the little glitches that were showing up in the pictures he took.

Still another guy was almost comical. He was a hematology researcher, i.e., he knew more about human blood than anyone else. He had a high-priced microscope that had a special light source. It was a glass tube that had been evacuated, and then refilled with a special combination of rare gases that each gave off a different color light when ionized. There were about a half dozen electrodes on

the tube that each had to see a different voltage. The poor guy had to spend 45 minutes any time he wanted to use the microscope balancing the voltages on these interactive electrodes . . . it was a touchy thing, I suspect. Once in a while, usually (as Murphy's law dictates) when he could ill afford the time, a power-line transient would commutate the tube, extinguishing the light. After cursing and yelling at his graduate research assistant (GRA), he would spend another half-hour to 45 minutes redoing the job. When we gave him an isolation transformer it solved both his technical problem and the GRA's blood pressure problem.

The standard for digital equipment today is to withstand the ANSI standard pulse of 20 µS and 2 kV. Beyond that point, we have to provide a little magic of our own.

120/240 VOLT ELECTRICAL SYSTEM

The standard 120/240 volt electrical system is shown in Figure 7.1 (this is for the United States and Canada; other countries will differ). Transformer *T1* is the "pole-pig" transformer

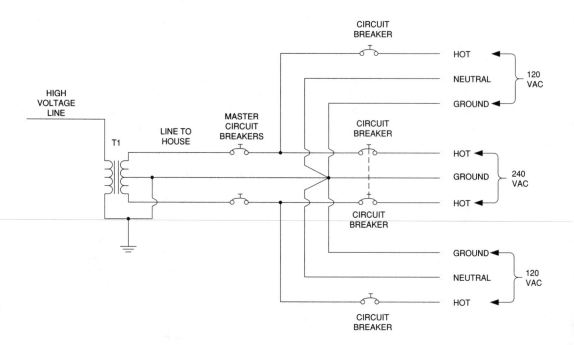

Fig. 7.1   Residential 120/240 volt AC power system.

outside of your house. The purpose of that transformer is to reduce the high voltage that the power is transmitted under to 240-volts AC center-tapped. Three wires are brought into the house where they encounter a pair of master circuit breakers (which may be breakers or fuses). From there the circuits branch out.

The 240-VAC circuit is used to run heavy appliances (dryers, stoves, air conditioners, etc). It is operated *across* the two hot lines and has its own set of circuit breakers. A ground wire is provided to keep the circuit safe.

One 120-VAC circuit is provided from each side of the transformer, making two independent circuits. Each circuit has its own circuit breaker.

The independent circuits are used separately, but there is some interaction through the neutral line. Normally, one expects to see the lights on a line *decrease* in brilliance when a large load (e.g., a compressor) comes on. But what happens when the neutral is open? In that case, the lights will become more *brilliant* when the heavy load comes on. This occurs because the 120-VAC lines are not loaded the same, and as a result of the neutral being open. The high drain of the compressor starting up is in series with the low drain of the light bulb, making for a very unbalanced situation.

NOISE

Noise can occur on the electrical system whenever there is sparking or any type of truncated waveform in use. Sparking can exist because of loose tie wires or other hardware in the high-voltage end of the circuit. We also see sparking on the 120/240-VAC side of the transformer due to electrical motor commutators, switches, and so forth. Sparking also occurs when there is a fault on the system.

The use of truncated waveforms occurs in TRIAC or SCR circuits when not all of the AC waveform is used. Figure 7.2 shows a truncated AC waveform. The harmonics generated by this scheme are tremendous. Recall that the sine wave is pure; all other wave-

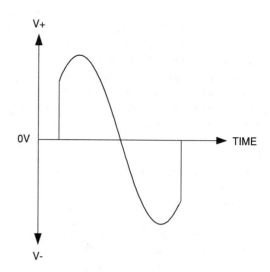

Fig. 7.2   *Truncated waveforms raise havoc with systems.*

forms have significant harmonics. Therefore, when not all of an AC waveform is used, harmonics are generated. Light dimmers are examples of such devices.

REGULATORY/LEGAL ISSUES

Power lines are regulated as incidental radiators under Part 15 of the Federal Communications Commission (FCC) Rules and Regulations. This means that they will incidentally generate EMI, rather than produce it as a normal part of their operation. The Rules and Regulations regarding incidental radiators state that the device shall:

1. Use good engineering practices

2. Not cause harmful interference

The operator of an incidental device shall cease operating upon notification by the FCC, and shall promptly take steps to ensure that operation of the device does not cause interference. That sounds like strong language, but there is a hidden argument. The argument hinges on the definition of "harmful interference." What is harmful in one case might not be harmful in another. As a result, there are no absolute limits that the power company must meet. And electrical power is necessary, so there is a built-in bias against turning it off. The FCC uses cost, the number of people involved, severity of interference, and a host of other factors to determine whether or not to get involved. In general, it is wise to try troubleshooting the problem yourself before involving the FCC.

CORONA AND SPARK

The interference caused by power lines can be due to corona or sparking. Corona is ". . . a partial breakdown of the air that surrounds an electrical element such as a conductor, hardware, or insulator." A corona discharge is often visible as a pale blue light around the conductor. A voltage gradient must exist between two different points such as the conductor and ground. As a result, you will see corona discharge around 7.5-kV lines, but they are more common with 230-kV lines.

The frequency components of corona discharge EMI are shown in Figure 7.3. Note that the two lines are present in the graph. The left side is for dry conditions while the right side is for wet conditions. This establishes a continuum between which the actual noise level will be found. When conditions are dry, the noise falls off rapidly above about 25 MHz, but when the conditions are wet the noise level is significant well into the VHF spectrum. It can affect up to the lower VHF TV channels, the FM BCB, and even some of the aviation band. The corona is generally restricted to 1,000 yards from the power line.

Sparks occur when there is a gap across which the AC energy can leap. The gap can be intentional (as in certain types of motor) or accidental. In any event, there must exist a voltage gradient high enough to cause the electrical breakdown of the air or other gas. Sparking usually occurs when the gap is 0.06 inches or less.

Because of the fact that we use 60 Hz in the United States and Canada, there will be 120 voltage peaks per second. This means that there will be 120 instances per second that are capable of causing noise peaks. This causes a characteristic "buzz" that can be heard at broadband frequencies to 1 or 1.5 GHz. Unlike corona noise, spark noise tends to die off when conditions are wet.

Both spark and corona noise is normally louder close to the power lines, which

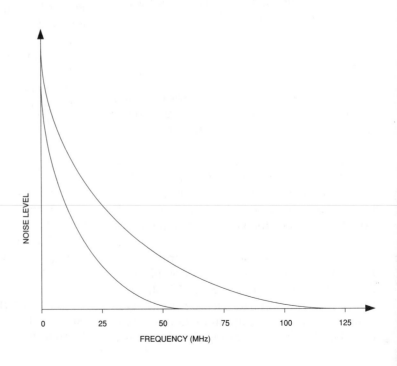

Fig. 7.3
Noise level vs frequency.

helps in locating it. Unfortunately, the noise will vary as one tracks along the line, *peaking* at the point where the noise occurs. This can cause you to misidentify points where the noise occurs.

SAFETY

Whether you are working with the high-voltage transmission lines or the 120/240 volts AC lines, there is high voltage present on the lines. It is essential that you operate in a manner that is consistent with safety when working on those lines. These lines can kill you . . . so don't take any chances.

LOCATING EMI SOURCES

Before attempting to find an EMI source outside the home or business, it is necessary to eliminate sources inside the building. Every device that has a motor in it is capable of generating EMI, so be careful. It is recommended that you read Chapter 11 before attempting to locate EMI.

FILTER SOLUTION

The filter solution works best when the source is at least partially shielded, but will work at least some most of the time. Figures 7.4A and B shows two circuits that can be used to filter AC noise carried on the AC power line. Interestingly enough, these filters can be used on either generators or victims of EMI.

Notice that each of these filters uses a special type of inductor. The inductor is designed as a common-mode coil having a single core for the two coils. This is not strictly necessary, but is a highly recommended form of construction.

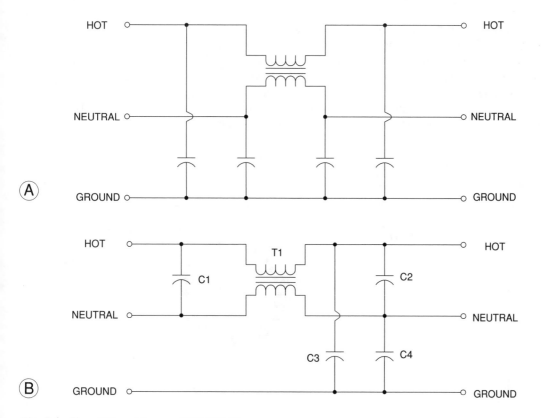

Fig. 7.4 *Two different forms of AC EMI filter.*

The capacitors selected are 0.01 to 0.1 μF of a type intended for use an the AC line. They will typically have a WVDC rating of 1,400 volts and will be rated for use in AC circuits. Lesser capacitors (such as 600 WVDC types) will not be able to withstand the peak and transient conditions.

Such filters should be installed as close as possible to the offending equipment. This eliminates the wiring to/from the sparking contacts as radiated sources of EMI. Alternatively, when protecting equipment from EMI due to power-line conduction, place the filter as close as possible to the point where the power line enters the cabinet of the equipment. There are EMI AC filters available that are built right into the AC plug assembly.

ELECTRIC MOTORS

There are two types of AC motor available: induction (Figure 7.5) and brush-commutator (Figure 7.6). Generally speaking, it is the brush-commutator motors that cause the most problems because of the sparking that occurs. Whether vacuum cleaners, electric driers, sewing machines, mixers, or power tools, the brush-commutator motor will cause problems. The fix is to add capacitors such as seen in Figure 7.6 and filtering. Attempt the capac-

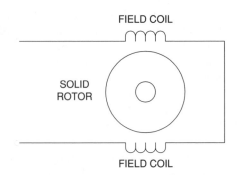

Fig. 7.5   *Induction motor.*

itors first, and then follow up with filters if that doesn't work.

COMMON-MODE FILTERING

Regardless of whether or not there is filtering or capacitors in the circuit, there should be a common-mode filter in the circuit. In fact, rolling up the power cord is the first thing that ought to be tried because it is nonintrusive. Figure 7.7 shows two versions of the common-mode choke installed on a TV set. It could just as easily be installed on any device that generates or is victimized by power-line EMI. In Figure 7.7A is a version using a linear rod, while Figure 7.7B is based on a toroidal core.

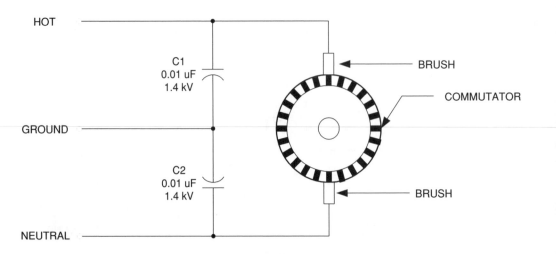

Fig. 7.6   *Brush-commutator motor.*

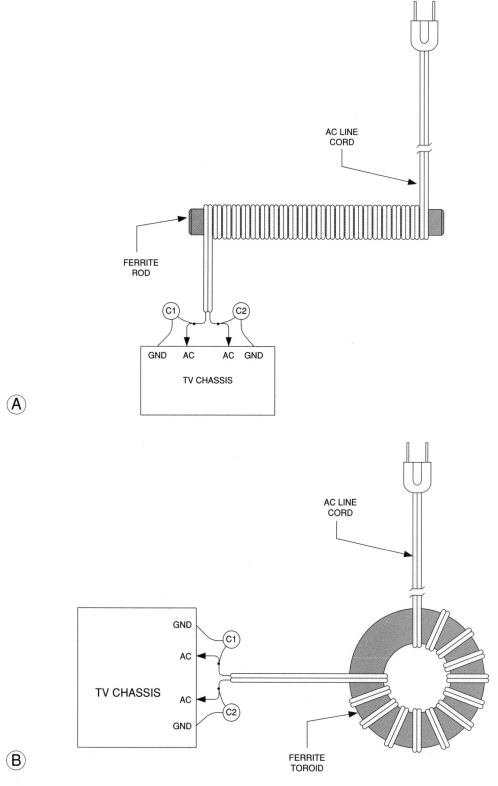

***Fig.* 7.7** *Two forms of common-mode choke: (A) ferrite rod, (B) toroid core.*

Chapter 8

Controlling Transmitter Spurious Emissions

Radio transmitters are a particular problem in dealing with EMI because they normally produce relatively large amounts of radio-frequency energy. If other devices respond to it inappropriately, then an EMI situation exists. It's not always the transmitter's fault, of course. The rule is simple: an electronic device must respond to the desired signals, and *not* respond to undesired signals. It's that second part that's a bit difficult. Many products are on the market that will respond inappropriately to a local transmitter.

TYPES OF TRANSMITTER

Before discussing how to deal with EMI from transmitters, let's first look at several different transmitter architectures. Figure 8.1 shows the simple master oscillator power amplifier (MOPA) design. This particular transmitter operates in the AM broadcast band (540 to 1,700 kHz) on a frequency of 780 kHz. The signal is generated at a low amplitude in a master oscillator (MO). In some older trans-

mitters the MO might be a variable-frequency oscillator (VFO), but for most the frequency will be controlled by a piezoelectric crystal (as shown).

The stability and operation of the MO are key to the performance of the transmitter. Of course, the short-term and long-term stability are set by the MO. In order to prevent problems due to the DC power supply, most designers will provide a special voltage regulator that serves only the oscillator. Even if other stages also have a voltage regulator, the MO will have its own.

The output of the MO is rarely sufficient to drive the final RF power amplifier, so there is usually some sort of buffer amplifier or driver amplifier between the output of the MO and the final amplifier. This stage does two things: (1) it builds up the MO output signal to the level required at the input of the final power amplifier, and (2) it provides a constant load to the oscillator. The latter is necessary because load variations translates into frequency variation in some oscillator circuits.

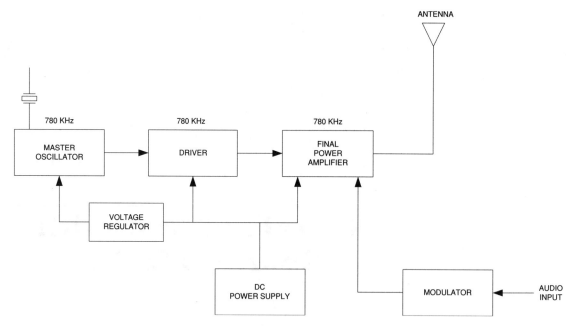

Fig. 8.1 *Master oscillator power amplifier (MOPA) type of transmitter.*

The final power amplifier is the work-horse of the transmitter. It takes the signal and boosts it to the level required for the application. Final power amplifiers may produce only a few hundred milliwatts, or a few megawatts, depending on the particular transmitter and its application. The final power amplifier usually includes some form of impedance matching to the load, and some form of tuning to prevent harmonics and other spurious emissions from getting to the outside world.

Transmitters usually have some means of imparting what is euphemistically called "intelligence" onto the RF carrier signal. Some form of modulator circuit is needed to accomplish this purpose. In the case shown in Figure 8.1 the audio modulator is applied to the final amplifier stage. This is called high-level amplitude modulation. It is also possible to modulate the driver stage.

A multiplier type transmitter design is shown in Figure 8.2. This circuit is simplified by not showing the modulator and DC power supply, even though they would, of

course, be present in an actual transmitter. This architecture was common on VHF/UHF transmitters at one time, and is still used on some. It allows a low-frequency signal to be multiplied into the VHF/UHF band where it can be transmitted. The benefit of this design is that good-quality crystals are easily available for low frequencies and become increasingly expensive and difficult for higher frequencies. At some point, the frequency is high enough that no crystal can be obtained.

Let's consider the example of Figure 8.2 in which the transmitter must output a signal on 152.155 MHz. The signal is generated in a master oscillator at 8.45306 MHz. The MO signal is applied successively to a series of frequency multipliers. These circuits are nonlinear, so a sine wave input signal will generate harmonics when processed in a nonlinear circuit. In the case of a doubler, the output is tuned to the second harmonic of the input frequency. In a tripler, the output is tuned to the third harmonic. In this particular case:

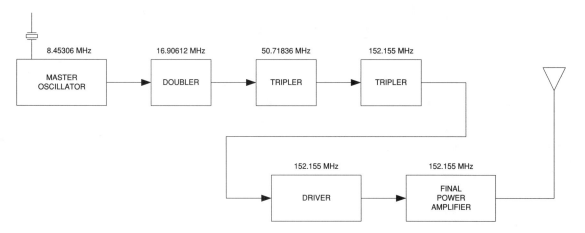

Fig. 8.2 *Frequency multiplier MOPA type of transmitter.*

Doubler output: 16.90612 MHz

Tripler 1 output: 50.71836 MHz

Tripler 2 output: 151.155 MHz

Once the multipliers have done their job, the 152.155-MHz signal can be applied to the driver and final RF power amplifier. The role of these stages is the same as in the MOPA design.

Multiplier transmitters are particularly troublesome from an EMI perspective because there is intentional generation of harmonics. A tuned circuit is used at each multiplier output to select only the desired harmonic. Unfortunately, no such circuit is, and at least some harmonic energy will pass through to the next stage. If the following stage is another multiplier, then the harmonic signal is also multiplied in that following stage. A real mess can result.

To add insult to injury, some designers, in an ill-conceived attempt to save money, will perform the multiplication function in either the driver or, worst of all, the final amplifier. The closer the multiplication process is to the output, the more likely it is that the spurious emission problem will arise. It is extremely easy on some models to tune the multiplier to the wrong harmonic, and if this happens in the final amplifier it will be radiated at a high

power level. It is highly recommended that a spectrum analyzer be used to tune up the multiplier style of radio transmitter.

A type of architecture that is increasingly popular is the mixer circuit depicted in Figure 8.3. This particular transmitter is a high-frequency (HF) single-sideband (SSB) transmitter. The SSB signal might be an upper sideband (USB) or lower sideband (LSB) signal. The LSB/USB signals are generated in a sideband generator consisting of the master oscillator, a balanced modulator, and a bandpass filter. The balanced modulator is used to suppress the RF carrier. Its output will be both upper and lower sidebands. The selection of LSB or USB is left to the bandpass filter.

In this particular type of design, which represents most SSB transmitters, the selection of LSB or USB is accomplished by selecting different MO crystals. The USB crystal places the carrier on the lower end of the filter bandpass characteristic, while the LSB places it at the upper end. One could also build an SSB transmitter with only a single crystal, and then use two filters (one each for LSB and USB). That approach, however, is costly and unwise.

Because of the filtering and other requirements of the SSB signal, it is advisable to generate the SSB signal on a single fre-

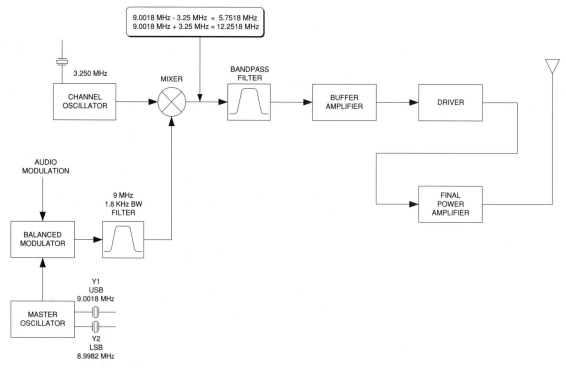

Fig. 8.3 *Single-sideband transmitter.*

quency. But because of its nature, the SSB signal cannot be translated by frequency multiplication. To translate the SSB generator output frequency (in this case 9 MHz), the signal is applied to an RF mixer circuit. The mixer combines the SSB signal with a signal from a second oscillator (called here the channel oscillator). In the case shown, the 9-MHz SSB signal is mixed with a 3.25-MHz channel oscillator signal to produce outputs at either the difference (5.7518 MHz) or sum (12.2518 MHz).

Following the mixer is a bandpass filter that selects either the sum or difference product, while rejecting the unwanted product. This signal is then applied to a buffer amplifier and/or driver amplifier, before being applied to the final RF power amplifier. In an SSB transmitter all stages following the SSB generator, except the mixer, must be linear. Thus, less efficient linear amplifier circuits are required for the buffer, driver, and final amplifier.

The existence of the mixer stage tells us that there is a possibility of products other than the desired products. The output spectrum of the mixer will be the familiar $mF_1 \pm nF_2$. In double balanced mixer (DBM) circuits, the two input signals are suppressed in the output (F_1 and F_2 are severely attenuated), leaving the products in which m and n are not 0 or both are not 1.

The problem is to reduce the unwanted mixer products as much as possible. The distant products are not particularly troublesome because they can be suppressed by filtering. But the third-order difference products ($2F_1 - F_2$ and $2F_2 - F_1$) are troublesome because they fall close to F_1 and F_2. These products may well be within the normal passband of the properly adjusted transmitter.

It is quite common today to find the master oscillator replaced with a frequency synthesizer. Figure 8.4 shows the basic

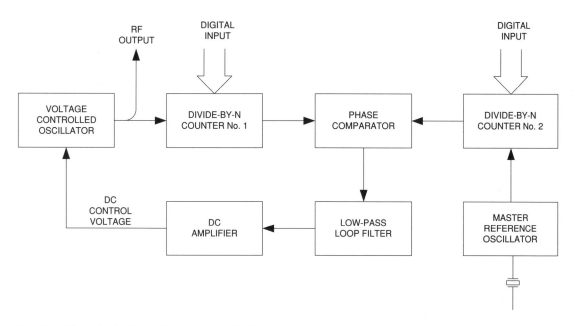

Fig. 8.4 Phase-locked loop frequency synthesizer.

phase-locked loop (PLL) circuit used in modern synthesizers. The actual RF output signal is produced in a voltage-controlled oscillator (VCO). The operating frequency of this type of oscillator is a function of a DC control voltage applied to its tuning input.

The PLL works by comparing the VCO output signal frequency to a fixed reference frequency, and then issuing a correction by changing the DC control voltage an amount proportional to the difference. The RF signal from the VCO is first reduced in frequency by a divide-by-N counter. A master reference oscillator (MRO) is used to control the process. The stability and accuracy of the overall PLL is controlled by the MRO. In some cases, a divide-by-N counter is used at the output of the MRO. Both the divided VCO signal and divided MRO signals are applied to a phase comparator circuit. This output of this circuit is filtered by the loop filter, and then scaled or level translated in the DC amplifier to the proper voltage. The voltage change at the output of the DC amplifier is designed to pull the VCO frequency in the direction that mini-

mizes the difference between the VCO and MRO frequencies (as divided). Selection of the channel frequency is done by controlling either or both divide-by-N counters' digital input.

OPERATING THE TRANSMITTER

When you operate a radio transmitter legally you are expected to transmit only on the assigned frequency, and none other. Ideally, when you generate a radio-frequency (RF) signal, only that frequency is created. And when the perfect ideal-but-never-achieved signal is modulated, the only new signal components that appear are those that are created by the modulation sidebands. Lots of luck. The real world can be a bit nastier. Let's take a look at some of the different forms of output signal normally found coming from a transmitter.

Figure 8.5 shows the amplitude-vs-frequency spectrum of a hypothetical transmitter. This type of display will be seen on

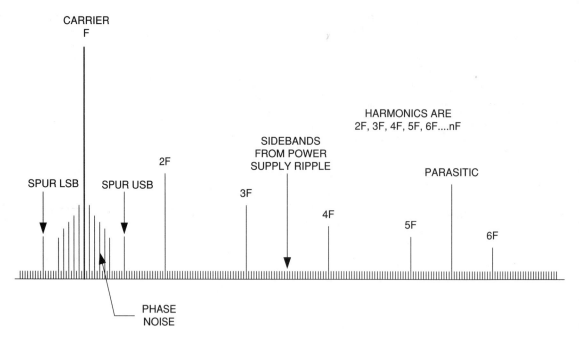

Fig. 8.5   *Output spectrum of a typical transmitter.*

an instrument called a spectrum analyzer (i.e., a frequency-swept receiver with its output connected to an oscilloscope that is swept with the same sawtooth as the receiver local oscillator).

The main signal is the carrier (*F*) and is the highest-amplitude "spike" in the display. We will consider only an unkeyed continuous wave signal because modulation sidebands would make a mess out of our clean little picture. All of the amplitudes in Figure 8.5 are normally measured in units called "dBc" for "decibels below the carrier." A signal that is −3 dBc, for example, would be 3 dB lower than the carrier, or about half the power of the carrier. For spurious outputs, the lower the level the better, so look for high negative dBc values (e.g., −60 dBc or more).

Direct Instability

If both the input and output ends of an RF amplifier are tuned (often the case), or if there is unexpected coupling between input and output causing feedback of the output signal, then the amplifier may oscillate on either the transmitter operating frequency or a nearby frequency. If the transmitter's "on frequency" output level does not drop to zero when the drive signal is reduced to zero, then suspect direct instability as a cause of oscillation. This is especially likely if the input and output are tuned to the same frequency, giving rise to what used to be called "tuned-grid–tuned-plate (TGTP) oscillators."

Phase Noise

Because the process of generating the single frequency *F* is not perfect, there will be a certain amount of noise energy surrounding the carrier. Some of these signals are caused by thermal noise in the circuit, as well as other sources. These signals tend to modulate the carrier creating the phase-noise sidebands shown in Figure 8.5.

Harmonics

Any complex waveform can be represented by a series of sine and cosine waves that make up its Fourier series or Fourier spectrum. If a transmitter produces a pure sine wave output signal, then only carrier frequency F will appear in the spectrum. But if the signal is distorted in any way, no matter how little, then harmonics appear. These signal components are integer multiples of the harmonic, so will appear at $2F$, $3F$, $4F$, . . . nF. For example, a 780-kHz AM broadcast band transmitter will have harmonics at 1,560 kHz, 2,340 kHz, 3,120 kHz, 3,900 kHz, and so forth. The specific harmonics and their relative amplitudes will differ from one case to another, depending upon the circuit and the cause of distortion.

Like all spurious emissions, the harmonics must be suppressed so that they do not cause interference to another service.

VHF/UHF Parasitics

A transmitter may be designed to operate at a relatively low frequency (medium-wave or high-frequency bands), but produce a large output signal in the VHF or UHF bands. The problem is due to stray capacitances and inductances in the circuit. Although the problem exists in transistor RF power amplifiers, it is most common in vacuum-tube RF amplifiers. Unfortunately, those amps that use tubes today are the really high-power units used in broadcast transmitters . . . so the problem is particularly severe because of the power levels.

Barkhausen's criteria for oscillation are (1) phase shift of 360 degrees at the frequency of oscillation, and (2) loop gain of one or more at that frequency. If the phase inversion of the amplifier device plus the frequency-selective phase shifts caused by the stray capacitances and inductances (including those inside components!) add to 360 degrees at any frequency where there is gain, then there will be an oscillation. Because strays are typically small, the oscil-

lation frequency tends to be in the VHF and up portions of the radio spectrum.

Power Supply Ripple

Transmitters like to work from direct current, but the power company supplies alternating current (AC) at a frequency of 60 Hz (some countries use 50 Hz). We use DC power supplies that convert AC to an impure form of DC called pulsating DC. This form of DC comes from the output of the rectifier and contains a *ripple factor* impurity at a frequency equal to the AC line frequency (60 Hz) in half-wave rectifiers or twice the AC line frequency (120 Hz). The ripple factor represents a small amplitude variation that tends to amplitude modulate the carrier. This produces a low-level "comb" spectrum with RF signals spaced every 120 Hz up and down the band.

Normally, this power supply artifact is not a problem, but if the DC power supply ripple filtering is ineffective, or if the application is particularly sensitive, then it will be heard. In some cases, such as the 400-Hz power supplies used in aircraft systems, or transmitters that use 5-kHz to 100-kHz switching power supplies, the problem can be much worse.

Low-Frequency Spurs

When an amplifier is misadjusted, or when an RF feedback path exists through the DC power supply (or other circuits), then there is a strong possibility of the amplifier oscillating at a low frequency (perhaps audio or below). These oscillations will amplitude modulate the RF signal, giving rise to a spurious emission. I've seen solid-state VHF RF power amplifiers break into low-frequency oscillation when either mistuned or biased incorrectly. If the low-frequency oscillation is due to DC power supply coupling, then use both a high-value electrolytic and low-value disk ceramic (or similar) capacitors in parallel for decoupling.

Frequency Halving

Solid-state bipolar transistor RF power amplifiers sometimes show an odd spurious emission in which a signal is produced at one-half of the carrier frequency. This phenomenon is seen when the input and output load and/or tuning conditions are such that the transistor operating parameters vary over cyclic excursions of the signal. Unfortunately, this effect is seen in a nonlinear situation, so odd multiples of the halving frequency occur. Suspect this to be the problem when a spurious emission occurs at 1.5F, because it could be the third harmonic of a halving situation.

Audio (and Other) Stage Oscillation

Few transmitters produce a single frequency with no modulation, so Figure 8.5 is rather simplistic in order to be able to illustrate the actual case. When the transmitter is modulated (AM, PM, FM, etc.), sidebands appear. Let's consider only the AM case for simplicity's sake. Let's say we have a 1-kHz audio sine wave tone modulating a 1,000 kHz (1 MHz) RF carrier. When the modulation occurs, a new set of sideband signals appear: the lower sideband (LSB) will appear at 1,000 kHz − 1 kHz = 999 kHz, and the upper sideband (USB) will appear at 1,000 kHz + 1 kHz = 1,001 kHz. In the case of a voice amplifier, the nominal range of audio frequencies is about 300 Hz to 3 kHz, so the normal speech sidebands will appear at ±3 kHz from the carrier, or in our 1,000-kHz case from 997 kHz to 1,003 kHz.

But what happens if the audio stages oscillate at a frequency higher than the audio range? The LSB/USB pairs will appear at those frequencies as well. I recall a VHF FM transmitter used in the 2-meter amateur radio band (144 to 148 MHz) that produced signals every 260 kHz up and down the band from the transmitter's nominal output frequency. The cause turned out to be "ultrasonic" oscillation of the FM reactance modulator stage. The manufacturer supplied a retrofit kit that provided better decoupling (capacitors and ferrite beads) and grounding of the circuit.

Once the oscillation ceased, the RF output was cleaned up.

Keep Barkhausen's criteria for oscillation in mind: Any time there is a frequency at which the loop again is greater than unity and the overall phase shift is 360 degrees, there will be an oscillation. This is true regardless of whether the stage is an audio amplifier, reactance modulator, or RF stage.

WHAT TO DO?

There are three basic strategies for reducing emissions to the level required by the Federal Communications Commission and good manners: (1) adjust (or repair) the transmitter correctly, (2) use shielding, and (3) filter the output of the transmitter.

The adjustment issue should go without saying, but apparently it is a problem. One trick that many transmitter operators pull is to either increase the drive to a final RF power amplifier to increase the output power, or peak the tuning for maximum output. This isn't always the smartest thing to do. Never operate the transmitter at levels above the manufacturer's recommendations. There are cases where tuning up the amplifier using a spectrum analyzer as well as an RF power meter will show that the increased power level apparent on the meter is due to the production of harmonics or other spurs and not the carrier.

At one time it was relatively common to see illegal operation of citizens-band transmitters. In the tube days, it was relatively easy to increase the RF output power from 4 watts on average to about 8 watts. Consider this situation: a 2:1 increase is only 3 dB, which is about half an S-unit on a distant receiver . . . or about half as much as the minimum discernible change. Yet, operating the transmitter that way not only doesn't produce the desired end result, it creates a distinct possibility of high harmonic or other spurious emissions!

When repairing a transmitter, use only parts that are recommended. This is especially true of capacitors and semiconductors. All capacitors exhibit a bit of stray inductance, as well as capacitance, and so will have a self-resonant frequency. If that frequency meets Barkhausen's rules, then oscillation will occur. In general, the problem comes from using a cheaper capacitor, or a different type of capacitor, from the one used originally. Also, be very wary of "replacement transistors" that are not really and truly exact replacements. These frequently cause either UHF or low-frequency oscillations.

Shielding is absolutely required in transmitters, especially higher power transmitters. Even low-power transmitters can cause spurious emission levels that will interfere with other services. Transmitters operated outside their cases, or with critical shields removed, are candidates for high radiation of spurs.

Even small shields are important. I recall one transmitter that had a large amount of AM splatter, and a very broad signal, as well as output components appearing up and down the band. The rig was a 300-watt AM HF band transmitter. The thing looked normal, but a photo of the transmitter in a service manual revealed a missing bit of metal on the master oscillator shielded housing. Someone had removed that little bit of sheet metal and allowed a slot to appear that admitted RF from the final to the oscillator housing. That feedback path proved critical. Restoring the shielding fixed the problem.

Filtering is shown in Figures 8.6 and 8.7. The filter is shown in Figure 8.6 at the output of the transmitter, in the transmission line to the antenna. Ham operators use low-pass filters for their HF transmitters because they are trying to protect VHF/UHF bands (especially TV channels). In other cases, a high-pass filter or bandpass filter may be used, depending on the frequencies that need protecting.

One of the best approaches to filtering is to use an absorptive filter as shown in Figure 8.7A. The absorptive filter has two filters, one high-pass and one low-pass, with the same cutoff frequency. Either filter can be used for the output, depending on the case. Let's consider the ham radio situation in

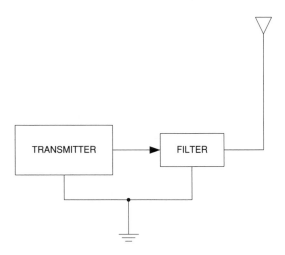

Fig. 8.6 *Filter used at the output of transmitter.*

which a high-frequency transmitter must protect TV channels in the VHF band. In that case, we will use the low-pass filter to feed the load ($R1$ represents the load, such as an antenna), and the high-pass filter will feed a nonradiating dummy load. The harmonics and parasitics, therefore, are absorbed in the dummy load, while the desired signal is output to the load.

In other cases, where the protected frequencies are below the transmitter frequency, then the roles of the high-pass and low-pass filters are reversed. $R2$ becomes the load and $R1$ becomes the dummy load.

Some absorptive filters will also place a wavetrap across the load in order to protect specific frequencies. In one version, there is a 40-MHz absorptive filter with a 56-MHz notch filter (i.e., series-tuned L-C circuit across the load). The design was published in *The ARRL RFI Book* for ham transmitters. I modeled the circuit on Electronics Workbench and produced the frequency response shown in Figure 8.7B. Note that the gain of the filter drops off starting just before 40 MHz (which is the −3 dB point), and there is a deep notch at the 56-MHz point. The design appears to be successful. In some cases, such as VHF or UHF communications systems, the series-tuned L-C network might be replaced with a cavity-tuned filter.

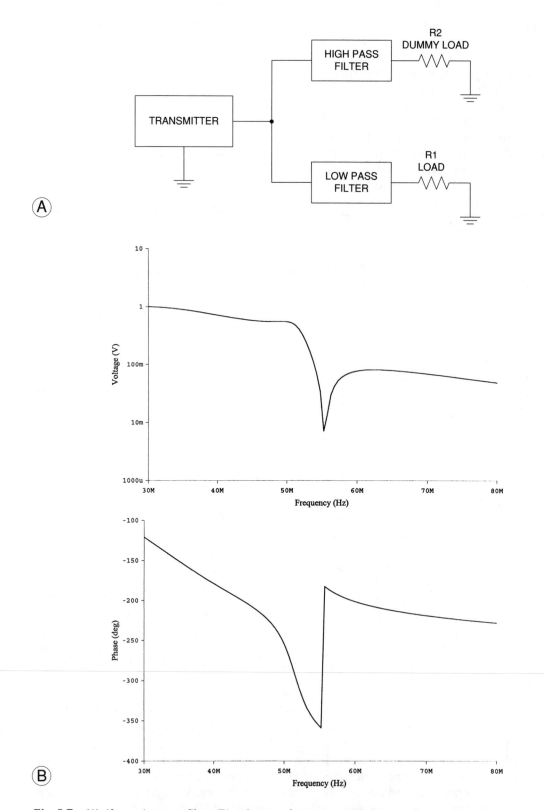

Fig. 8.7 *(A) Absorption type filter; (B) voltage vs frequency and phase vs frequency.*

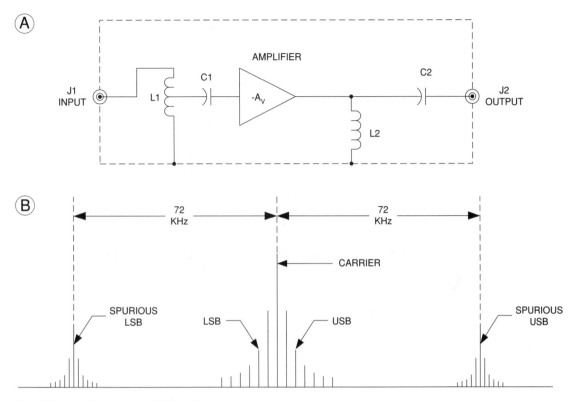

Fig. 8.8 *(A) Shielded amplifier; (B) spectrum.*

BE WARY!

One peculiar form of low-frequency oscillation occurs in supposedly broadband solid-state power amplifiers. In some units, a broadband toroid transformer couples input and output to the transistors of the power amplifier (Figure 8.8A). A DC blocking capacitor will be used to prevent bias from being shorted out through the transformer. Unfortunately, the inductance of the transformer and the capacitance of the coupling capacitor form a tuned resonant circuit. If both input and output are tuned to the same frequency, then a species of TGTP-like oscillator is formed. Such oscillations tend to occur in the 10- to 200-kHz range and give rise to spurious RF output sidebands (SPUR LSB and USB in Figure 8.5) spaced at that frequency from the carrier, *F*. This effect is shown exaggerated in Figure 8.8B.

TRANSMITTER TEST SETUP

Figure 8.9 shows a typical test setup for transmitters. For mobile, aviation, and marine units, a variable DC power supply is used, while base station or fixed units can be powered from a variable AC transformer such as a Variac.

The instrumentation setup will vary somewhat with each situation, but that shown in Figure 8.9 is representative. Of course, if no external RF power amplifier is used, one RF wattmeter can be deleted. An output RF wattmeter and dummy load are used regardless of the type of transmitter. An isolated coupler or sampling tee is used to take a small portion of the signal to the test instruments. A peak deviation meter is a staple of FM transmitter servicing, but would not be part of other transmitter setups. Sometimes, an oscilloscope is also used. In many cases, either a spectrum analyzer or a specialized communications service monitor (or "test set") is also used.

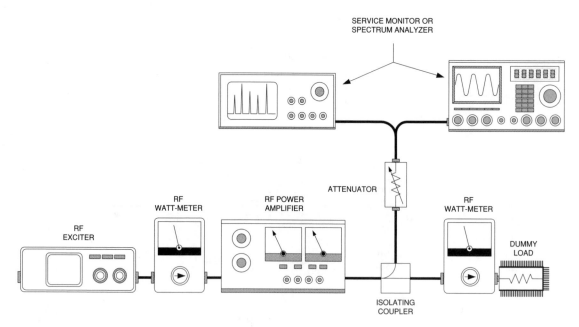

Fig. 8.9 *Test setup for transmitter–linear amplifier.*

THIRD-HARMONIC OR HIGHER

The third harmonic (or higher) can be eliminated using a scheme such as that shown in Figure 8.10. This is a stub method and is applied to the base of the antenna radiator element. It consists of two quarter-wavelength stubs, one in parallel

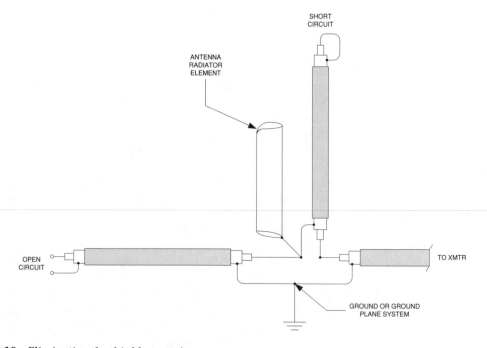

Fig. 8.10 *Eliminating the third harmonic.*

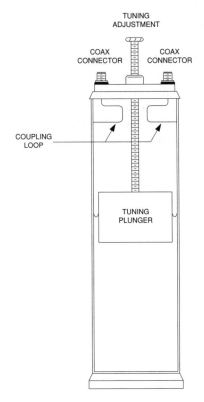

Fig. 8.11   Cavity resonator.

with the line and one in series. The open stub is placed in parallel with the line, while the shorted stub is placed in series with the line.

VHF AND UP TRANSMITTERS

At VHF and up the possibility exists to use high-Q tuned-cavity resonators to separate frequencies in duplex systems, or to eliminate specific nearby interference. Figure 8.11 shows the tuned cavity in sectioned view. It consists of a tuned cavity in which input and output coupling loops are placed. The "tuned" part comes from the fact that the dimensions of the cavity can be changed using the tuning plunger. It acts like a series resonant circuit.

Figure 8.12 shows an application of the tuned-cavity resonators in a repeater scheme. The circuit is called a duplexer. The repeater scheme means that the receiver and transmitter have to operate at the same time, making the EMI problem critical. Good reception requires that the noise of the transmitter at the

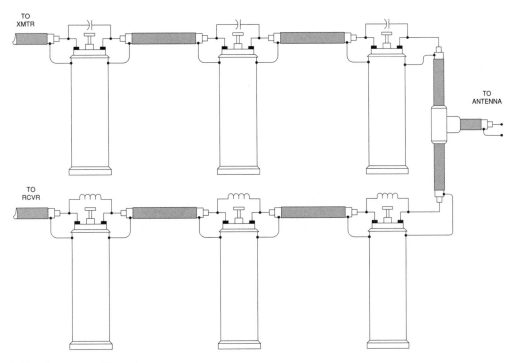

Fig. 8.12   Repeater with cavity resonators.

receiver frequency be attenuated. The transmitter side of the duplexer is tuned to the transmit frequency, and the receiver side is tuned to the receiver frequency.

Let's consider an example. The receiver has a sensitivity of –132 dBm. The transmitter has an output that is –80 dBc. At an output of 60 watts (+48 dBm) the noise level is –32 dBm. The duplexer must make up the difference between them, or about 132 dBm + (–32 dBm) = –100 dB. The receiver must make up the difference between the +48 dBm output of the transmitter and the overload point. This can be as high as +10 dBm, and as low as –36 dBm. We will assume the worst case and provide attenuation of –36 dBm – 48 dBm = –84 dBm. In each case, there are three tuned cavity resonators required.

Chapter 9

Telephones and EMI

This chapter will deal with telephone and modem EMI. Interference to telephone receivers is almost always the fault of the receiver or its wiring. That means the users or owners of the telephone system are at fault and will have to bear the expense and effort to clear up the EMI problem. That said, it doesn't do you a lot of good as the technician assigned to clean up the interference to put the job off on the user!

THE FEDERAL COMMUNICATIONS COMMISSION

Basically, the FCC has no responsibility in the matter of EMI to telephones. Ever since deregulation, there has been no reason why the FCC is interested. The FCC regards this as a matter between the parties involved and does not get involved.

The FCC does, however, receive a large number of complaints about EMI to phone systems. As a result, their Compliance & Information Bureau publishes a booklet that may be of some help: "What to Do If You Hear Radio Communications on Your Telephone" (Bulletin CIB-10).

THE TELEPHONE COMPANY

The local telephone company may view EMI problems as a matter of interest to those involved, and of no particular interest to themselves. Their responsibility ends with the "drop" at the service entrance to the business or home.

Ever since deregulation the local phone companies have not had any interest beyond the drop. However, some do offer wiring contracts where the user pays a monthly fee on their telephone bill to have the phone company be responsible for the wiring. Most such contracts will exempt the telephone receiver itself. Nonetheless, these contracts are worth exploring if there is an EMI problem on the telephone lines. Most such EMI problems are line problems, especially in the 500-kHz to 40-MHz region. Otherwise, if you are going to involve the telephone company, discuss who is going to pay them.

THE RADIO OWNER

FCC regulations state that radio signals must be free of those characteristics that interfere with other services. That means radio and television services, not telephones. The radio owner and operator have no responsibility toward the telephone owner. That said, it is incumbent on the responsible operator of the equipment to try to provide the solution to the interference problem. But what that means is open to discussion. It may mean, for example, handing out filters. Or it may mean handing out good advice. The issue of personal diplomacy is of primary concern here, because the person being interfered with feels aggrieved.

THE TELEPHONE MANUFACTURER

Ever since the deregulation of the telephone industry, there has been a tremendous number of firms providing telephones to the U.S. marketplace. Some of them deal with EMI complaints effectively; others do not. You would think that manufacturers and importers of telephone equipment would be interested in the interference-free use of that equipment. Such is not always the case. In fact, several importers don't have the technical capacity to be of much help to the user. Some manufacturers, however, are a little better and will even supply filters for those customers who experience interference.

THE TELEPHONE OWNER

The responsibility for EMI proofing telephone equipment falls squarely on the end user of the equipment. Regardless of whether the interference is due to conducted pickup

in the wiring, or due to direct pickup in the instrument itself, there is nothing the radio operator can do to eliminate the problem.

The problem is due to poor design of the telephone equipment selected by the consumer. The use of a filter should be explored. Failing that, a consumer may opt for one of the EMI-proof telephones that are offered.

A problem is that the interfering station appears to be doing something to the telephone. That sometimes produces a bitter consumer who demands that the problem be cleared up on the transmitter end. A little "personal diplomacy" will go a long way toward settling that issue, I suspect.

TECHNICAL ISSUES

All that has gone before in this chapter assumes that someone will get the job of defining a solution to the EMI problem. In this section, we will discuss the telephone wiring system and what can be done about it.

Figure 9.1 shows two forms of wiring that may be present in a home or business. Figure 9.1A shows the parallel wiring scheme, and Figure 9.1B shows the loop series wiring scheme. In the parallel scheme there are as many wires from the junction block where the phone company's interest terminates as there are telephones. In the loop series wiring scheme there is one pair of wires leaving the junction block, and the telephone instruments are daisy-chain wired from them. In truth, there might be a situation where both methods are used (which would reflect a user wiring scheme) . . . as shown in Figure 9.1C!

There will be a fused lightning arrestor present at the drop. This lightning arrestor might be corroded, or it may be nonlinear. In case of corrosion the nonlinearity may be due to oxides or bimetallism causing the device to act like a diode. In either case, the nonlinearity will act like a diode and cause rectification of RF energy. This in turn causes the EMI.

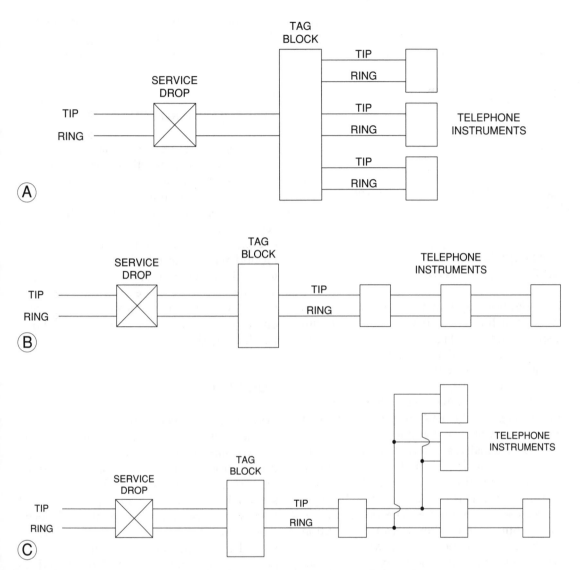

Fig. 9.1 *(A) Parallel wiring system; (B) loop series wiring system; (C) typical wiring system.*

TWISTED PAIR, FLAT (PARALLEL), AND SHIELDED WIRING

Twisted pair wiring is superior to flat wire where EMI is concerned. This is because twisted pairs are self-shielding, whereas flat wire is not. Twisted pair wiring can pass through noisy environments and be EMI free.

Unfortunately, most telephone wiring in the United States appears to be flat wire, eliminating the self-shielding aspects of twisted pair wiring. If you have a situation where there is a lot of flat wire involved, then it would be prudent to change it. For really difficult cases, use telephone company Category 3 or Category 5 wiring, even if the cost is high. Alternatively, you can use shielded wire yourself. The shield will protect against the EMI, but is terribly expensive.

It is good practice to ground unused wires or pairs of wires in a bundle. If the unused wires are grounded at the service entrance to the house, then there will be a decrease in susceptibility to EMI.

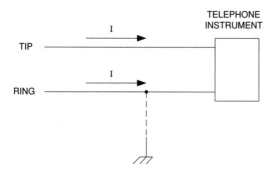

Fig. 9.2   *Grounded telephone system could spell trouble.*

COMMON MODE VS DIFFERENTIAL MODE

The telephone system is normally common mode, which is to say that it is balanced with respect to ground. Figure 9.2 shows this condition. When common-mode integrity is maintained the currents (I) flowing as a result of RF will be equal, nulling out at the instrument. It is not guaranteed that common-mode lines will be EMI free, but there is a higher probability of that condition existing.

Occasionally, a differential-mode condition exists. This is an unbalanced condition. This is illustrated by the grounded condition in Figure 9.2. This mode can exist because staples used to mount the wire break one insulator but not the other, because insulation breaks down, or because of wiring errors. When the differential mode exists, the telephone will still operate as before, but the susceptibility to EMI will be tremendously increased.

RESONANCES

The telephone wiring in a home or business acts like a random length or even long-wire antenna to signals in the 500-kHz to approximately 200-MHz region. In that case, there will be resonance effects, i.e., those lengths that are integer multiples of quarter wavelength. The effect appears to peak in the 1.6-

to 15-MHz region, but is present throughout the region of the spectrum cited above. At these resonant points there is an increased probability of EMI problems.

TELEPHONE GROUND

The telephone system is grounded at the service entrance and no other point. If grounds exist elsewhere, they must be dealt with appropriately . . . remove them. A properly installed telephone ground will either use a separate ground rod or be tied to the power company's ground. But there is a kind of ground where the telephone ground is tied to a cold water pipe at one end of a house, and the cold water pipe is in turn grounded to the AC power mains ground at the service entrance. This is a bad practice and should be eliminated.

CORROSION

Corrosion can occur on wires in the telephone system. This occurs because of years of exposure in damp spaces. Corrosion can cause noise on the line (hissing and frying, or pops), as well as making the line more susceptible to EMI. The susceptibility to EMI occurs because corrosion makes a decent diode, especially when two metals are involved in the junction (e.g., tinned vs copper). Wiring that is corroded should be replaced, or at least the corroded portions should be eliminated and soldered.

SUBSTANDARD WIRING

There is a substantial possibility of encountering substandard wiring in troubleshooting EMI-to-telephone problems. This is especially likely now that consumers are doing their own wiring. Whenever substandard wiring is encountered, it should be replaced before any attempts, other that simple filtering, are made to eliminate the EMI. This is especially true if one of the wires is shorted to ground or there is corrosion present.

TELEPHONE CLASSIFICATION

Telephones are classified according to type: plain old telephones or multifeatured. The plain old telephone is relatively free of EMI, but not entirely. This is because the features of the other type require electronic circuitry in which there may be semiconductor junctions that can rectify RF. Generally speaking, simple filtering (see below) deals with plain old telephone EMI. Filtering alone may or may not help the multifunction EMI problem. In those cases, shielding may be in order.

TELEPHONE REGISTRATION NUMBERS

Many telephones are sold in the United States with no importer or manufacturer listed. There will be a manufacturer's registration number present. In case you want to contact the manufacturer, the registration number can be interpreted for you by the Federal Communications Commission (Manager Part 68 Rules, 2000 M Street NW, Washington, DC, 20554). Once it is determined that the telephone itself is at fault, contact the manufacturer for information on fixes.

CAPACITORS

The capacitor is a differential-mode filter of sorts and can be used to eliminate EMI. Use a 0.001- to 0.01-μF, 1,000 WVDC disk ceramic unit (the voltage rating is needed because the ring voltage can exceed 100 volts).

Unfortunately, this simple fix won't always work. At 3,000 Hz (the maximum frequency response of a telephone) the impedance of a 0.01-μF capacitor is about 5,300 ohms. Several capacitors will produce a value of impedance that may disrupt service, making it low volume.

The capacitor will not work in the case of EMI to a high-speed modem. Even low-value capacitors will exhibit enough phase shift to interrupt the operation of these devices.

COMMON-MODE RF CHOKES

If the telephone is responding to common-mode signals, then a common-mode choke may do wonders for the EMI problem. The simplest form of choke is the ferrite, such as shown in Figure 9.3. Use the type 73, 75, or 77 ferrite material for the lower HF range, and type 43 ferrite for the VHF range. This type of choke is placed as close to the instrument as possible.

Figure 9.4 shows a common-mode choke that is useful for EMI up to about 30 MHz or so. Figure 9.4A shows the schematic symbol; Figure 9.4B shows the physical form of an actual choke. The turns of the common-mode choke are wound in the *bifilar* manner, i.e., they are wound together. This is done either by paralleling the wires, or by twisting them together prior to winding. If you plan to build

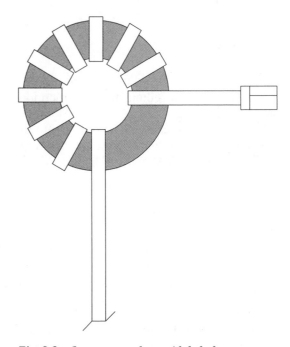

Fig. 9.3 *Common-mode toroidal choke.*

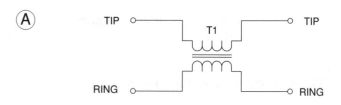

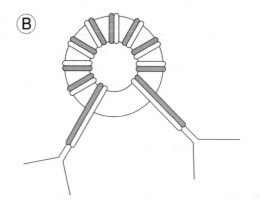

Fig. 9.4
Toroidal choke: (A) Schematic; (B) wiring.

your own, use about 28 turns of No. 30 AWG wire on a 0.50-inch form.

FILTERING

There is a basic fact of life in telephone EMI problems: filtering eliminates a tremendous amount of it! The instrument can't respond to EMI if the RF that causes EMI can't get to the instrument. It's as simple as that! A sample telephone filter is shown in Figure 9.5, al-

though you will probably want to buy a filter rather than make one. It is most useful when the filter is placed as close to the affected instrument as possible.

When dealing with commercial filters, it may be necessary to work with several products, especially if the frequency of the offending station is not well established. The filters are made for different bands, and that factor alone makes them useful for different types of EMI.

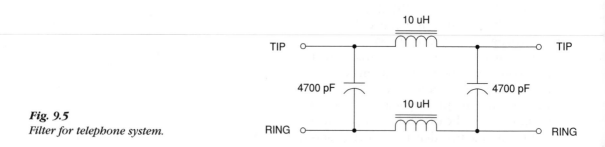

Fig. 9.5
Filter for telephone system.

FEDERAL COMMUNICATIONS COMMISSION
COMPLIANCE & INFORMATION BUREAU

Telephone Interference
Bulletin CIB-10 August 1995

WHAT TO DO IF YOU HEAR RADIO COMMUNICATIONS ON YOUR TELEPHONE

Interference occurs when your telephone instrument fails to "block out" a nearby radio communication. Potential interference problems begin when the telephone is built at the factory. All telephones contain electronic components that are sensitive to radio. If the manufacturer does not build in interference protection, these components may react to nearby radio communications.

Telephones with more features contain more electronic components and need greater interference protection. If you own an unprotected telephone, as the radio environment around you changes, you may sometimes hear unwanted radio communications. Presently, only a few telephones sold in the United States have built-in interference protection. Thus, hearing radio through your telephone is a sign that your phone lacks adequate interference protection. This is a technical problem, not a law enforcement problem. It is not a sign that the radio communication is not authorized, or that the radio transmitter is illegal.

Because interference problems begin at the factory, you should send your complaint to the manufacturer who built your telephone. A sample complaint letter is provided here. You can also stop interference by using a specially designed "radio-proof" telephone, available by mail order. A recent FCC study found that these telephones, which have built-in interference protection, are a very effective remedy. A list of Radio-Proof telephones is provided here.

Interference problems in telephones can sometimes be stopped or greatly reduced with a radio filter. Install this filter at the back of the telephone, on the line cord, and/or at the telephone wall jack. Radio filters are available at local phone product stores and by mail order. (See attached list, Radio Interference Filters.) A list of Radio Interference filters is provided here.

To get started, follow these steps: If you have several telephones, or accessories such as answering machines, un-plug all of them. Then plug each unit back in, one at a time, at one of your wall jacks. Listen for the radio communication. If you hear interference through only one telephone (or only when the answering machine is plugged in), then the problem is in that unit. Contact the manufacturer of that unit for help. Alternatively, simply stop using that unit, replace it with a radio-proof model, or install a radio filter. (NOTE: Only a very small percentage of interference problems occur in the outside telephone lines. Your local telephone company can check for this type of problem.)

Next, it's important to follow through and contact the manufacturer. Telephone manufacturers need to know if consumers are unhappy about a product's failure to block out radio communications. Also, the manufacturer knows the design of the telephone and may recommend remedies for that particular phone.

To file a complaint, write a letter to the manufacturer. To help the manufacturer select the right remedy, be sure to provide all the information in the sample, including the type of radio communication that the telephone equipment is receiving. You can identify the type of radio communication by listening to it. There are three common types:

1. AM/FM broadcast radio stations—Music or continuous talk distinguishes this type of radio communication. The station identifies itself by its call letters at or near the top of each hour.

2. Citizen's Band (CB) radio operators— These radio operators use nicknames

or "handles" to identify themselves on the radio. Usually, the CB operator's voice is clearly heard. You may also hear sound effects or other noises.

3. Amateur ("ham") radio operators— Amateur radio operators are licensed by the FCC. They use call letters to identify their communications. The amateur's voice can be heard but may be garbled or distorted.

Cordless telephones are low-power radio transmitters/receivers. They are highly sensitive to electrical noise, radio interference, and the communications of other nearby cordless phones. Contact the manufacturer for help in stopping interference to your cordless telephone.

Final note: Current FCC regulations do not address how well a telephone blocks out radio communications. At present, FCC service consists of the self-help information contained in this bulletin. A partial list of radio-proof telephones and radio filters is also attached.

The FCC strongly encourages manufacturers to include interference protection in their telephones as a benefit to consumers. The telephone manufacturing industry has begun to develop voluntary standards for interference protection. The FCC will continue regular meetings with manufacturers and will closely track the effectiveness of their voluntary efforts. If you are not satisfied with the manufacturer's response, contact the Electronic Industries Association, 2500 Wilson Blvd., Arlington, Virginia 22201, phone: (703) 907-7500.

SOURCES OF RADIO-PROOF TELEPHONES AND RADIO FILTERS FOR TELEPHONES

The lists below show companies that sell radio-proof telephones and radio interference filters. If you would like to try a radio-proof telephone or radio interference filter, make sure that you can return it for a refund, and keep the purchase receipt.

RADIO-PROOF TELEPHONES:

TCE LABORATORIES, INC.
2365 Waterfront Park Drive
Canyon Lake, TX 78133
(830) 899-4575

NOTES:
Desk and wall models available. Will do custom orders for multiple-line phones, speaker phones, answering machines, etc. Advertises 30-day money-back guarantee.

PRO DISTRIBUTORS
2811 74th Street, Suite B
Lubbock, TX 79423
(800) 658-2027

NOTES:
Desk and wall models available. Advertises 30-day money-back guarantee.

RADIO INTERFERENCE FILTERS:

AT&T
(800) 222-3111

NOTES:
Also available at AT&T and GTE Phone Center stores.

COILCRAFT
1102 Silver Lake Road
Cary, IL 60013
(800) 322-2645

NOTES:
Filters for computers and printers also available.

ENGINEERING CONSULTING
583 Candlewood Street
Brea, CA 92621
(714) 671-2009

NOTES:
Also available filters for 2-line telephones.

INDUSTRIAL COMMUNICATIONS
ENGINEERS (ICE)
P. O. BOX 18495
Indianapolis, IN 46218-0495
(317) 545-5412

NOTES:
Also available hard-wired filter for wall-mount telephone.

K-COM
P.O. Box 82
Randolph, OH 44265
(330) 325-2110

NOTES:
Also available filters for 2-line telephones.

KEPTEL, INC.
56 Park Road
Tinton Falls, NJ 07724
(732) 389-8800

KILO-TEC
P. O. Box 10
Oak View, CA 93022
(805) 646-9645

OPTO-TECH INDUSTRIES
P.O. Box 13330
Fort Pierce, FL 34979
(800) 334-6786, or (407) 468-6032

RADIO SHACK (ARCHER)
Available at nearest Radio Shack store.
Catalog #273-104.

SNC MANUFACTURING
101 W. Waukau Avenue
Oshkosh, WI 54901-7299
(800) 558-3325, or (414) 231-7370

SOUTHWESTERN BELL FREEDOM PHONE ACCESSORIES
7486 Shadeland Station Way
Indianapolis, IN 46256
(800) 255-8480, or (317) 841-8642

TCE LABORATORIES, INC.
2365 Waterfront Park Drive.
Canyon Lake, TX 78133
(830) 899-4575

NOTES:
Also available filters for 2-line telephones.

Chapter 10

Noise Cancellation Bridges

Whether you operate a radio receiver, or some piece of scientific or medical instrumentation, noise interferes with acquiring desired signals. Noise is bad. After all, radio reception and other forms of signals acquisition are essentially a game of signal-to-noise ratio (SNR). The actual values of the desired signal and noise signal are not nearly as important as their ratio. If the signal is not significantly stronger than the noise, then it will not be properly detected.

Getting rid of noise battering a signal is a major chore. Although there are a number of different techniques for overcoming noise, the method described herein can be called the "invert and obliterate" approach. This same idea was used in a popular novel in which a cranky inventor created a dynamic stealth concept by placing antennas all over an aircraft to receive radar signals, invert them, and then retransmit them 180 degrees out of phase with the incident wave . . . thereby causing cancellation. The idea is also used in actual (not fictional) noise abatement systems in which microphones and loudspeakers are used to retransmit room noises 180 degrees out of phase with the incoming. According to reports I've heard, remarkable

reductions in local noise are possible, although the technique tends to fall down over large areas.

Figure 10.1 shows the basic problem and its solution (cast in terms of radio reception). The signal from the main antenna is a mixture of the desired signal and a locally generated noise signal. This noise signal is usually generated by the 60-Hz alternating current (AC) power lines, or machinery and appliances operating from the 60-Hz AC lines. The noise signal is not confined to 60 Hz, but will extend into the VHF region because of harmonic content. The noise spikes will appear every 60 Hz from the fundamental frequency up to about 200 MHz or so, although the harmonics become weaker and weaker at progressively higher frequencies. But in the VLF bands (where they are often overwhelming), AM broadcast band (AM BCB), and medium-wave shortwave bands, the noise signal can be tremendous. It will therefore cause a huge amount of interference.

The solution (also shown in Figure 10.1) is to invert the noise signal and combine it with the signal from the main antenna. When the phase-inverted noise signal combines with the noise signal riding on the

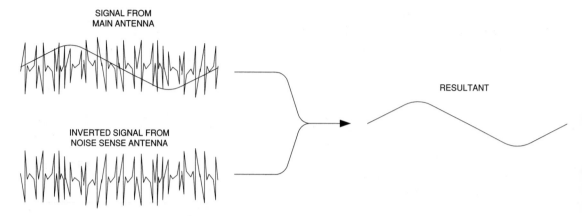

Fig. 10.1 *The problem and the solution: combine a phase inverted noise signal with the regular noise-plagued signal.*

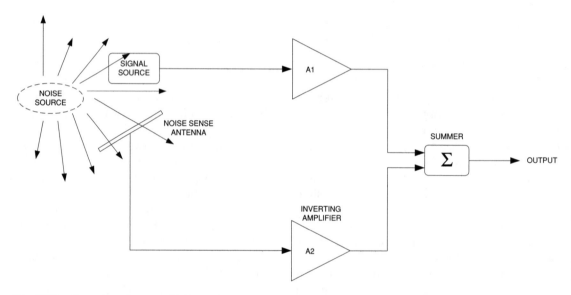

Fig. 10.2 *Generic noise cancellation scheme.*

main signal, the result is cancellation of the noise signal, leaving the resultant main signal. What is needed is a noise sense antenna, a means for inverting the noise signal, and a summing circuit.

Figure 10.2 shows a generic case that serves to illustrate the method for both radio reception and other forms of instrumentation. The signal source will be the main antenna in the case of radio reception. But in the case of scientific instruments it might be some sort of sensor. In a medical case, it could be a human patient with either sensors

or electrodes attached (as in an electrocardiogram). The noise source is the local AC power lines or machinery that radiates a signal of some sort. If the noise signal is picked up by the noise source (or its connecting wires), then it will travel through amplifier A1 and cause interference. But the signal can also be picked up by a small sense antenna and fed to an inverting amplifier (A2). By definition, an inverting amplifier shifts the phase of the input signal 180 degrees, so when the inverted noise signal is applied to the summer it will cancel the noise component of the main signal. It might be necessary to provide some amplitude control in order to not replace the main noise signal with a new noise signal from A2.

The case of a radio receiver system is shown in Figure 10.3. The phase inversion and summation functions of Figure 10.2 are performed in a special noise cancellation bridge circuit. The main antenna is the antenna that is normally used with the receiver. It might be a dipole, vertical, beam, or just a random length of wire strung between two trees.

The noise sense antenna is optimized for pick-up of the noise source signal. One VLF radioscience observer told me via e-mail that he uses a 36-inch whip antenna mounted on his roof as the noise sense antenna. In some shortwave situations the sense antenna is a 10- to 30-foot length of antenna wire running parallel to the power lines that are creating the noise.

> ### CAUTION
> *Under no circumstances should you allow the sense antenna to touch the AC power lines, even if it breaks and whips around in the wind. In the case of VHF noise reception the sense antenna might be a two- or three-element beam (Yagi or Quad) aimed at the noise source. Other combinations are also possible, I presume.*

One goal of the sense antenna is to make it highly sensitive to the local noise field, while being a lot less sensitive to the desired signal than the main antenna. Although in purist terms both noise and desired signals appear in both antennas, the idea is to maximize the noise signal and minimize the "desired" signal in the sense antenna, and do the opposite in the main antenna.

In the system of Figure 10.3, the noise sense signal and main signal are combined in

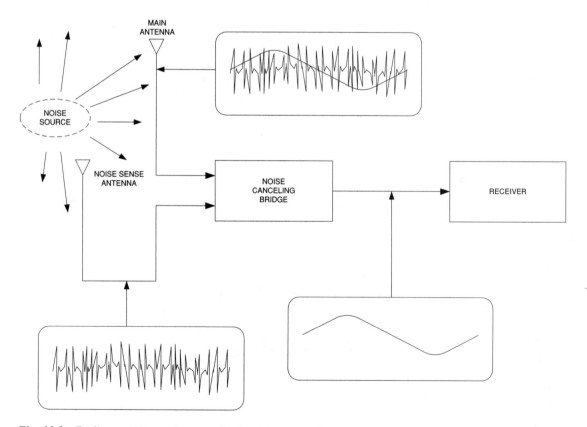

Fig. 10.3 *Radio reception noise cancelling system.*

a noise-cancelling bridge (NCB). The output of the NCB is a cleaned up version of the antenna signal, with greatly improved SNR.

The design problems that must be overcome in producing the NCB are easy to see. First, it must either invert or provide other means for producing a 180-degree phase shift of the noise signal. It must also account for amplitude differences so that the inverted noise signal exactly cancels the noise component of the main signal. If the amplitudes are not matched, then either some of the original noise component will remain, or the excess amplitude of the inverted noise signal will transfer to the signal and become interference in its own right. The noise signal inversion can be accomplished by transformers, bridge circuits, RLC phase shift networks, or delay lines.

A SIMPLE BRIDGE CIRCUIT

Figure 10.4 shows a simple bridge circuit. I've used it at VLF on radioscience observing receivers, and others have used it on VHF receivers. The bridge consists of two transformers (T1 and T2). Transformer T1 is trifilar wound, i.e., it has three identical windings interwound with each other in the manner of Figure 10.5. The black "phasing dots" or

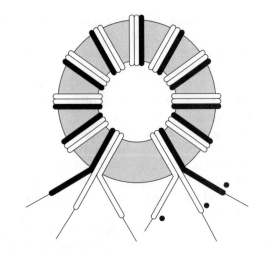

Fig. 10.5   Trifilar wound toroid transformer.

"sense dots" indicate one end of the windings and will be used for wiring T1 into the circuit of Figure 10.2.

Winding the toroid exactly as shown in Figure 10.5 is a difficult task, so you might want to consider an alternative method. Select three lengths of enameled wire (#18 AWG through #26 AWG can be used, but all wires should be the same size). In order to keep them straight in my mind as I work them, I select three different insulation colors from my wire rack.

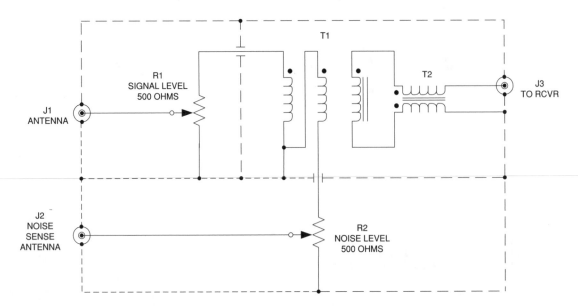

Fig. 10.4   Simple transformer-based noise-cancelling bridge.

Tie all of them together at one end, and insert that end into the chuck of a hand drill. I usually fasten the other ends into a bench vise and back off until the wires between the vise and drill are about straight (more or less). Turn on the drill on a slow speed (slightly squeeze the trigger on variable-speed drills), and let the drill twist them together. Keep this process going, being careful to not kink the wire (which happens easily!), until there are 8 to 16 twists per inch (not critical).

CAUTION

Wear protective goggles or safety glasses when doing this job. I once let the drill speed get too high, the wire broke, and I received a nasty lashing to the face ... which could've damaged my eye except for the glasses.

Once the three-wire composite wire is formed, it can be wound onto the toroid form as if it were one wire. Before winding, however, separate the ends a bit, scrape off enough insulation to attach an ohmmeter probe, and measure both the continuity of each wire and whether or not any two are shorted together. If the wires are wound too tightly, it's possible to break one wire, or breach the integrity of the insulation.

Note in Figure 10.4 the way transformer T1 is wired. The main antenna signal from J1 is connected to the dotted end of one wind-ing, while the sense antenna signal (J2) is ap-plied to the undotted end of another of the three windings. These signals are transferred to the third winding, but because of their rel-ative phasing (due to how they are connect-ed, dotted or undotted) they will be 180 degrees out of phase. The desired signal, however, appears only in the J1 port, and so will not be phase inverted.

The composite output of T1, i.e., the noise plus desired signal and the inverted noise signal, is applied to transformer T2. This trans-former is inserted into the line as a common-mode choke and so will perform the actual cancellation of the inverted and noninverted noise signal components. Transformer T2 is built exactly like T1, but is bifilar (two wind-ings) instead of trifilar.

Signal amplitudes from the two differ-ent antennas are controlled by a pair of 500-ohm potentiometers. The pots selected for R1 and R2 should be noninductive (i.e., car-bon or metal film, but *not* wirewound). In other words, rather ordinary potentiometers will work nicely. The wipers of both poten-tiometers are connected to their respective antenna jacks. The two ends are connected to T1 and ground, respectively.

Note that this circuit is not just built into a shielded box, but also in separate shielded compartments. Figure 10.6 shows a suitable form of building the circuits. A compartment-ed box such as one made by SESCOM is used to hold the bridge. Small grommets mounted on

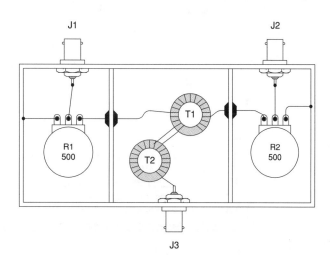

Fig. 10.6
Shielded construction.

the internal shield partitions are used to pass wires from one compartment to the other.

For VLF through shortwave, transformer T1 is wound with 16 turns of enameled wire, and T2 is wound with 18 turns. Both can be wound on 1/2-inch cores (T-50-xx or FT-50-xx), but it will be easier to use slightly larger forms such as FT-68-xx, T-68-xx, FT-82-xx, and T-82-xx. Ferrite cores (FT-nn-xx) should be used in the AM BCB and below, while powered iron (T-nn-xx) can be used in the medium wave and shortwave bands.

Recommended ferrite types for VLF through the AM BCB include FT-82-75 and FT-82-77; medium-wave units can be made using FT-82-61; and VHF units can be made using FT-67 or FT-68 (or their -50 and -68 equivalents). If powered iron cores (T-nn-xx) are used, then select T-80-26 (YEL/WHT) for VLF, T-80-15 (RED/WHT) for AM BCB and

low medium wave, either T-80-2 (RED) or T-80-6 (YEL) for medium wave to shortwave, and T-80-12 (GRN/WHT) for VHF. Again, their -50 and -68 equivalents are also usable. Some experimentation might be needed in specific cases depending on the local noise problem.

Figure 10.7 shows a version of the noise-cancellation bridge circuit made popular by William Orr (W6SAI) and William R. Nelson (WA6FQG) for amateur radio use (*Interference Handbook*, RAC Publications, P.O. Box 2013, Lakewood NJ, 08701). It is built on the same principles as Figure 10.4, but includes an L-C phase shift network consisting of L1, C1, and C2. The values are:

> T1: 16 turns, #18 AWG trifilar wound on Amidon FT-82-61 core
>
> T2: 18 turns, #18 AWG, bifilar wound on Amidon FT-82-61 core

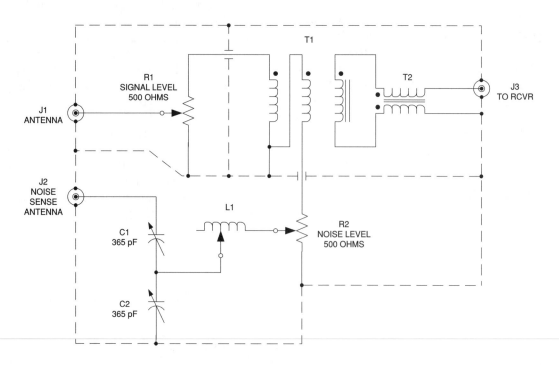

T1: 16 turns, #18 AWG Trifilar wound on Amidon FT-82-61 core
T2: 18 turns, #18 AWG, Bifilar wound on Amidon FT-82-61 core
L1: 32 uH, 45 turns, #22 enameled wire, 1-inch diameter, wire-dia. spaced
R1,R2: 500-ohms, non-inductive
C1,C2: 365 pF trimmer capacitors

Fig. 10.7 *Shortwave/HF variable noise bridge.*

L1: 32 μH, 45 turns, #22 enameled wire, 1-inch diameter, wire-diameter spaced

R1,R2: 500 ohms, noninductive linear taper potentiometer

C1,C2: 365-pF capacitors

The coil L1 should be wound with either enameled wire or noninsulated solid wire so that it can be tapped.

To adjust this bridge, C1, C2, and the tape on L1 should be adjusted iteratively until the lowest possible noise signal is achieved. To do this trick, it is usually necessary to set R1 and R2 to a low setting, but not so low that both the noise and the signal disappear.

A DIFFERENT BRIDGE

A somewhat different approach to the bridge concept is shown in Figure 10.8.

L1A: Twelve turns #22 AWG solid bare wire, 1-inch diameter, wound over 2-inch length

L1B: Five turns spaced one diameter apart, #22 AWG solid bare wire, wound over center of L1A (a layer of insulating black electrical tape must separate the two coils)

The potentiometers are 500-ohm, linear tape, noninductive pots of the type also specified for Figure 10.4. This bridge is tricky to balance as it involves the interaction of R1, R2, C1, C2, C3, L1A, and L1B. In some cases, one or both potentiometers must be shorted out to allow signal to pass unimpeded. In other cases, some value of R1 or R2 may be needed to balance amplitudes. Adjust all components iteratively until the best signal-to-noise ratio is obtained.

Parts can be a little difficult to obtain for RF projects, especially the capacitors. Ocean State Electronics [6 Industrial Drive, P.O. Box 1458, Westerly, RI, 02891, 1-401-596-3080 or FAX 1-401-596-3590] stocks both new and used variable capacitors, as well as various inductors, toroid cores, and other items of interest.

CONCLUSION

Radiated noise can be one of the most intractable electromagnetic interference (EMI) problems. These bridges are not a "silver bullet" by any means, but they will perform sufficient noise reduction to make a significant difference in the signal-to-noise ratio . . . and that's what actually counts.

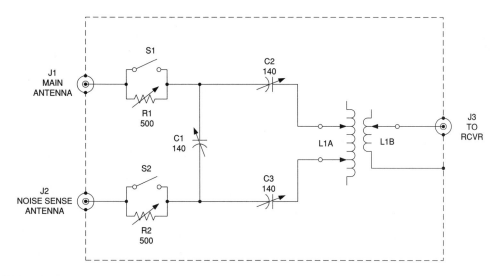

Fig. 10.8 *Alternate noise bridge circuit.*

Chapter 11

Locating EMI Sources

The first job in solving an EMI problem is to locate the source. Noise can be emitted from any of a wide variety of sources. Some are obvious (such as a radio transmitter), while others are not so obvious (such as dimmer switches in the lighting system). In this chapter we will look at some approaches to finding the sources, including both simple sleuthing tools and more formal radio direction finding (RDF) methods.

RF SLEUTHING TOOLS

The correct tool for finding RF sources is anything that will permit you to unambiguously determine where the radiation is coming from. We will take a look at several low-cost possibilities.

If you are looking for a source that is out in the neighborhood somewhere, such as a loose power line or a malfunctioning electrical system in another building, then an ordinary solid-state portable radio may do the trick. Open the radio and locate the loopstick antenna. If you rotate the radio through an arc where the loopstick is first broadside to the arriving signal, and then perpendicular to

the signal, you will notice a tremendous reduction in signal. Ferrite loopsticks are extremely sensitive to direction of arrival, with a sharp null occurring off the ends.

By noting the direction in which the null occurs (Figure 11.1), you find the line of direction to/from the signal source. It is not unambiguous, however. To find the actual direction, from the two possibilities that are 180 degrees opposed, move along the line and note in which direction the signal increases. Except in a very few cases where reflections occur, the signal will become stronger as you move toward the source.

The standard medium-wave and short-wave loop antenna used by a lot of radio enthusiasts is also useful for finding RF emitting sources. A square loop 30 to 90 cm on a side is relatively easy to construct and forms a very directional antenna. Indeed, these antennas are commonly used in radio direction finding. Such loop antennas offer a null when pointed broadside to the signal direction, and a peak when orthogonal to the signal direction.

A portable radio with an S-meter will work wonders in this respect. An ad-hoc S-meter can be formed using the circuit in Figure 11.2. This circuit plugs into the ear-

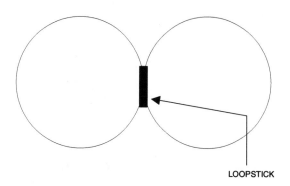

LOOPSTICK

Fig. 11.1 *Directional pattern of a ferrite loopstick antenna.*

BCBs). The HF antenna in those receivers is a telescoping whip. A loopstick sensor can be fashioned in the manner of Figure 11.3. An 18-cm (7-inch) ferrite rod is wound with about 20 turns of fine enameled wire. This wire is connected to a coaxial cable, which is in turn connected to the external antenna terminals of the receiver (if one exists). If there are no external antenna terminals, a small coil of several turns wrapped around the telescoping antenna will couple signal to the radio. It's a good idea to keep the antenna at minimum height in order to minimize pickup from sources other than the loopstick. More sensitivity can be had if a resonant loopstick is available, but those also restrict the frequency band.

A Rif Sniffer that is popular with mobile radio installers is shown in Figure 11.4. Although fancy commercial models exist for a price, the basic form of Figure 11.4 can be

phone jack of the receiver. The received audio is rectified by D1/D2 (a voltage doubler), and then applied to a DC current meter.

If the noise peaks in the HF shortwave bands, then the radio's loopstick is of little use (it only works on the MW and LF AM

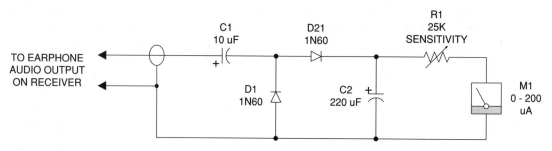

Fig. 11.2 *"Audio S-meters" for use on earphone-equipped receivers.*

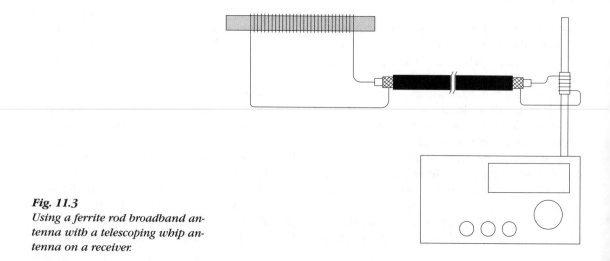

Fig. 11.3
Using a ferrite rod broadband antenna with a telescoping whip antenna on a receiver.

TAPE OR INSULATING CAP

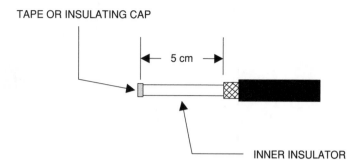

5 cm

INNER INSULATOR

Fig. 11.4
Simple coax sniffer.

made with a length of coaxial cable connected to the receiver antenna terminals. The shield is cut back a distance of 5 to 7 cm, and then removed. The inner conductor, covered with the inner insulator, becomes the sensor for finding RF sources such as ignition noise radiators.

In most cases, some kind of insulating cap is placed over the end of the inner conductor to keep it from contacting voltage sources as it is used to probe for RF. An accidental contact with the 12-volt battery of a vehicle can destroy the input coil on the receiver.

A "gimmick coil" sensor is shown in Figure 11.5. The sensor in this case is a solenoid-wound inductor connected to a length of coaxial cable. The coil has a diameter of

2.5 to 5 cm, and a length that is at least its own diameter. The coil consists of 2 to 10 turns of wire, depending on frequency (the higher the frequency, the fewer the turns required). The other end of the coaxial cable is connected to a receiver. When the coil is brought nearer the noise source, the signal level in the receiver will get higher.

A single-turn gimmick is shown in Figure 11.6. This sensor consists of a single loop, approximately 10 cm in diameter, connected to coaxial cable. It can be used well into the low end of the VHF spectrum, as well as at HF. The loop is made of small-diameter copper tubing, heavy brass wire or rod, or heavy-gauge solid copper wire (#10 or lower). The loop has some directivity, so it can be used to ferret out extremely localized sources.

A fault with the single-turn loop is that it is somewhat sensitive to magnetic fields, although not so much as some of the other forms. Figures 11.7A and 11.7B show dual-loop sensors that are less sensitive to magnetic field pick-up. In Figure 11.7A the loops of the sensor are rectangular and cross in the center. The feedline (coaxial cable) is connected to a break in one of the two coils.

COIL

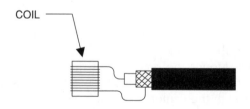

Fig. 11.5 *"Gimmick" coil sniffer.*

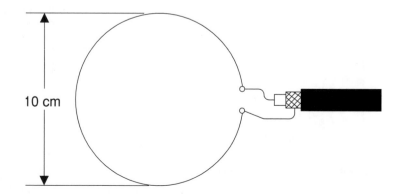

10 cm

Fig. 11.6
Single-turn loop sniffer.

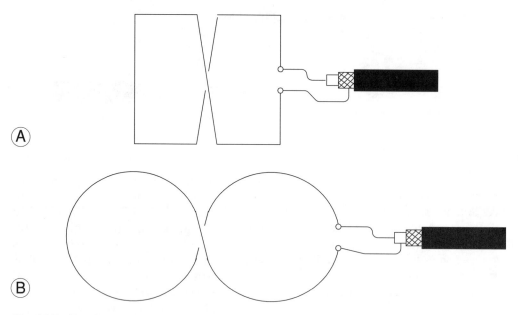

Ⓐ

Ⓑ

Fig. 11.7 *Two forms of anti-magnetic-field sniffer.*

The version shown in Figure 11.7B is circular. One version that I built was made of #8 AWG solid copper wire formed around a John McCann Irish Oatmeal tin can. Getting the coils about the right size and reasonably circular is relatively easy given a proper former. The exact size of the coils in Figures 11.7A and 11.7B is not terribly important. The coil should not be too large, however, or it will be less local and may become a bit ambiguous in locating some sources.

Figure 11.8 shows a method for sensing RF flowing in a conductor. This method is used as the sensor in a lot of ham radio RF power meters and VSWR meters. The conductor is passed through a toroid coil form, essentially acting as a one-turn primary winding of the transformer. The "secondary" winding is made of #22 to #30 wire wound over the toroid. About 6 to 20 turns are used, depending on frequency.

One popular method for constructing the sensor in Figure 11.8 is to order a toroid core that has an inside diameter a little less than the outside diameter of a rubber grommet. Mount the grommet in the center hole of the toroid core, and then pass the wire through the center hole of the grommet. The

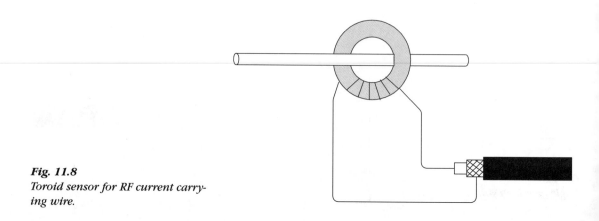

Fig. 11.8
Toroid sensor for RF current carrying wire.

two ends of the secondary winding are connected to the inner conductor and shield of the coaxial cable to the receiver.

RF DETECTORS

The RF output of the sensor coils can be routed to a receiver, and for low-level signals may well have to be so treated. For higher power sources, however, an RF detector probe is used. Figures 11.9 and 11.10 show two forms of suitable RF detector probe.

The RF detector in Figure 11.9 is passive, i.e., it has no amplification. It can be used around transmitters and other RF power sources. The input from the sensor is applied to C1, a small value capacitor, and then is rectified by the 1N60 diode. The 1N60 is an old germanium-type diode and is used in preference to silicon diodes because it has a lower junction potential (so is more sensitive). The junction potential of Ge devices is 0.2 to 0.3 volts, and for silicon it is 0.6 to 0.7 volts. The pulsating DC from the rectifier is filtered by capacitor C2. Resistor R1 forms a load for the diode and is not optional.

An amplifier version of the RF detector circuit is shown in Figure 11.10. In this version the same RF detector circuit is used, but it is preceded by a 15- to 20-dB gain amplifier. In this particular circuit the amplifier is a Mini-Circuits MAR-1 device. It produces gain from near-DC to about 1,000 MHz. Other devices in the same series will work to 2,000 MHz, and in the related ERA-x series up to 8,000 MHz. Clearly, any of these devices is well suited to the needs of most readers. The

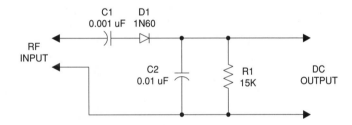

Fig. 11.9
RF detector converts RF signal to DC.

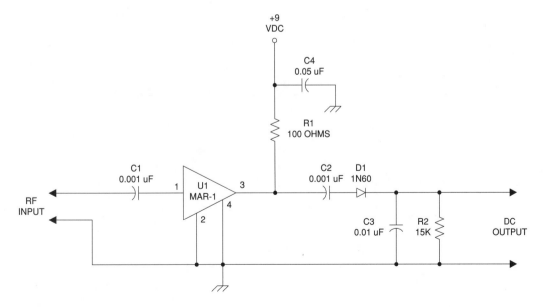

Fig. 11.10 *Amplified RF detector.*

cost of the MAR-1 device is very low (in the United States it is about $3 in unit quantities).

RADIO DIRECTION FINDING

Radio direction finding (RDF) is the art of locating a radio station or RF noise emitter by using a directional radio antenna and receiver. When someone wants to locate a signal source that is transmitting, they will use radio direction finders to triangulate the position (Figure 11.11). If they find the bearing from two sites only, then they will locate the station at the intersection. However, there is a fair degree of error and ambiguity in the measurement. As a result, radio direction finders typically use three or more (hence "triangulate") sites, as shown in Figure 11.11. Each receiving site that can find the bearing to the station reduces the overall error.

At one time aviators and seamen relied extensively on radio direction finding. It is said that the Japanese air fleet that attacked Pearl Harbor, Hawaii, on December 7, 1941, homed in on a Honolulu AM radio station.

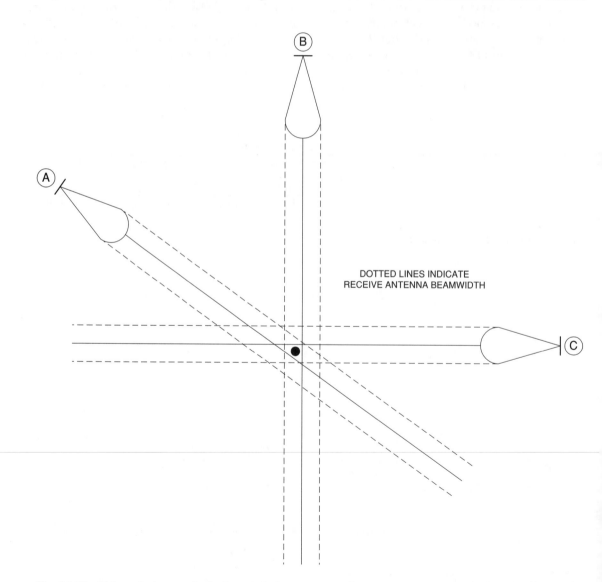

DOTTED LINES INDICATE
RECEIVE ANTENNA BEAMWIDTH

Fig. 11.11   *Triangulation method of precisely locating an emitter.*

During the 1950s and early 1960s AM radios came with two little "circled triangle" marks at the 640-kHz and 1,040-kHz points on the dial. These were the "CONELRAD" frequencies that you could tune to in case of a nuclear attack. All other radio stations were off the air except the CONELRAD stations. The enemy was prevented from using these frequencies for RDF because the system used several stations that transmitted in a rapidly rotating pattern. No one station was on the air long enough to allow a "fix." The result was a wavering sound to the CONELRAD station (which we heard during tests) that would confuse any dirty, smelly bad guy who tried to RDF his way into our cities with a load of nukes.

Radio direction finding received based on the AM BCB and shortwave bands looks a bit like Figure 11.12. A receiver with an S-meter (which measures signal strength) is equipped with a rotatable ferrite loopstick antenna to form the RDF unit. A degree scale around the perimeter of the antenna base could be oriented toward north so that the bearing could be read. A vertical whip is used as a "sense antenna" to unidirectionalize the loopstick.

Loopstick antennas have a "figure-8" reception pattern (Figure 11.13A) with the maxima parallel to the loopstick rod and the minima off the ends of the rod. When the antenna is pointed at the signal, maximum reception strength is achieved. Unfortunately, the maxima are so broad that it is virtually impossible to find the true point on the compass dial where the signal peaks. The peak is too shallow for that purpose. Fortunately, the minima are very sharp. You can get a good fix on the direction of the signal by pointing the minima toward the station. This point is found by rotating the antenna until the audio

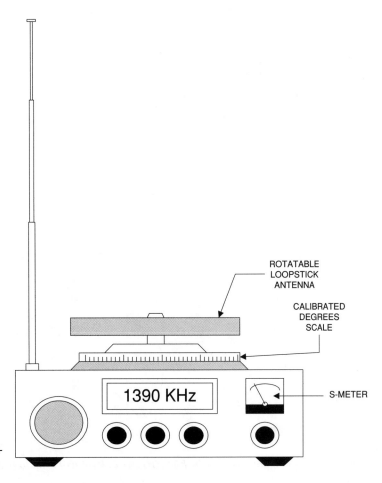

ROTATABLE
LOOPSTICK
ANTENNA

CALIBRATED
DEGREES
SCALE

1390 KHz

S-METER

Fig. 11.12 *Typical AM BCB/SW radio direction finder receiver.*

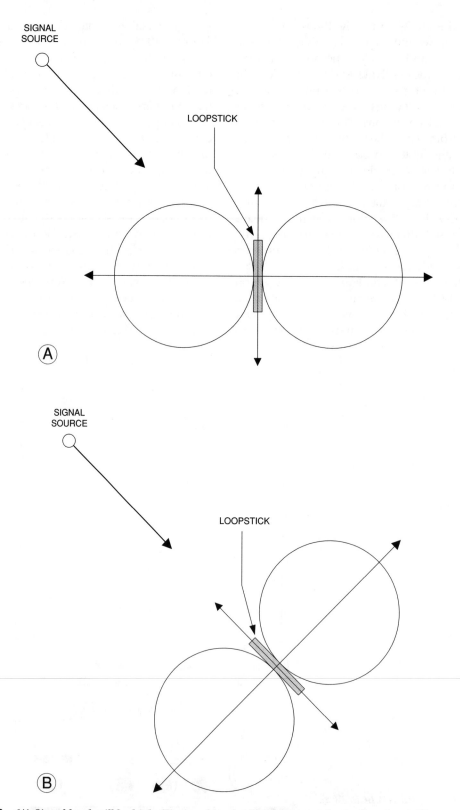

Fig. 11.13 *(A) Signal level will be high; (B) signal level will drop to nearly zero "on the null."*

goes to zip or the S-meter dips to a minimum (Figure 11.13B).

The loopstick is a really neat way to do RDF . . . except for one little problem: the darn thing is bidirectional. There are two minima because, after all, the pattern is a figure-8. You will get exactly the same response from placing either minima in the direction of the station. As a result, the unassisted loopstick can only show you a line along which the radio station is located—it can't tell you which direction it is. Sometimes this doesn't matter. If you know the station is in a certain city, and that you are generally south of the city and can distinguish the general direction from other clues, then the line of minima of the loopstick will refine that information. A compass helps, of course. Shortly we will take a look at an impromptu radio direction finder using a portable radio.

The solution to the ambiguity problem is to add a sense antenna to the loopstick (Figure 11.14A). The sense antenna is an omnidirectional vertical whip, and its signal is

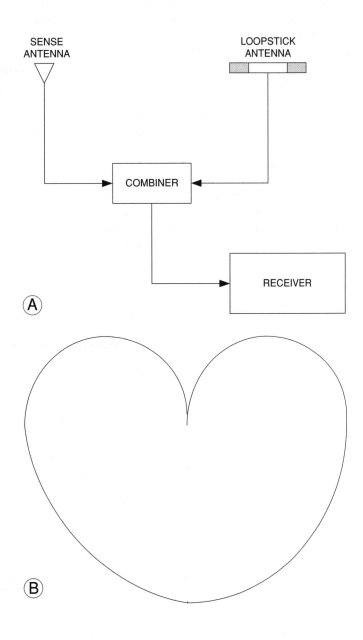

Fig. 11.14
(A) Use of a combiner network and sense antenna; (B) cardioid antenna pattern.

combined with that of the loopstick in an R-C phasing circuit. When the two patterns are combined, the resultant pattern will resemble Figure 11.14B. This pattern is called a cardioid because of the "heart" shape it exhibits. This pattern has only one null, so it resolves the ambiguity of the loopstick used alone.

FIELD IMPROVISATION

Suppose you don't have a decent RDF receiver, but you need to find the source of an EMI emitter nonetheless. The answer to your direction finding problem might be the little portable AM BCB radio (Figure 11.15) that you brought along for company.

Open the back of the radio and find the loopstick antenna. You will need to know which axis it lies along. In the radio shown in Figure 11.15 the loopstick is along the top of the radio from left to right. In other radios it is vertical from top to bottom. Once you know the direction, you can tune in a known AM station and orient the radio until you find a null. Your compass can give you the bearing. If you know the approximate location of the station or noise source, then you can re-verse the compass direction from it and mark the line on the topo map. Of course, it's still a bidirectional indication, so all you know is the line along which you are working.

To make the receiver better for noise hunting you might want to add either an external audio S-meter, an AC voltmeter (the AC section of a VOM or digital multimeter will work, but those with analog meter readouts are easier to use). The meter is connected to the earphone jack of the receiver. If your receiver does not have one, then you can add a 3/16-inch jack to the radio and connect it across the loudspeaker connections.

REGULAR LOOP ANTENNAS

Regular wire loop antennas (Figure 11.16) are also used for radio direction finding. In fact, in some cases the regular loop is preferred over the loopstick. The regular loop antenna may be square (as shown), circular, or any other regular "n-gon" (e.g., hexagon), although for practical reasons the square is easier to build. The loop has pretty decent inductance even with only a few turns. One loop I built was 24 inches square ("A" in

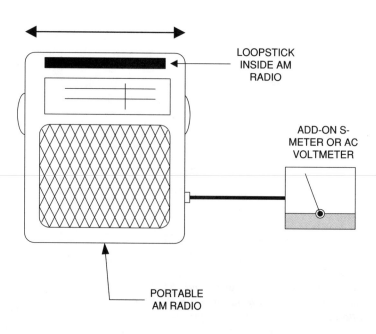

Fig. 11.15
Portable AM BCB receiver adapted for use as a noise sensor.

LOOPSTICK
INSIDE AM
RADIO

ADD-ON S-
METER OR AC
VOLTMETER

PORTABLE
AM RADIO

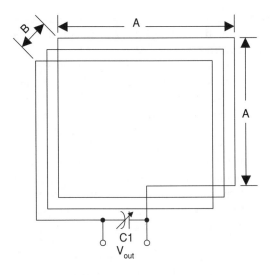

Fig. 11.16 *Loop antenna.*

voltage will be lower, but it will not require tuning.

When you use a regular loop antenna be aware that the antenna has a figure-8 pattern like the loopstick, but it is oriented 90 degrees out of phase with the loopstick antenna. In the regular loop those minima (nulls) are perpendicular to the plane of the loop, while the maxima are off the sides. In the antenna of Figure 11.16 the minima are in and out of the page, while the maxima are left and right (or top and bottom).

SENSE ANTENNA CIRCUIT

Figure 11.17 shows a method for summing together the signals from an RDF antenna (such as a loop) and a sense antenna. The two terminals of the loop are connected to the primary of an RF transformer. This primary (L1A) is center-tapped and the center-tap is grounded. The secondary of the transformer (L1B) is resonated by a variable capacitor, C1. The dots on the transformer coils indicate the 90-degree phase points.

Figure 11.16) and had, if I recall correctly, about 10 turns of wire spaced over a 1-inch width ("B" in Figure 11.16). It resonated to the AM BCB with a standard 365-pF "broadcast" variable capacitor. If you want the loop to be broadband, then delete C1. The output

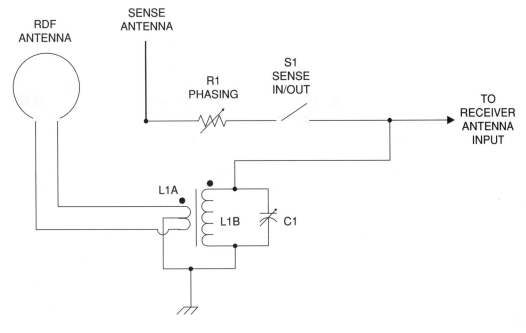

Fig. 11.17 *Sense antenna combiner circuit.*

The top of L1B is connected to the sense circuit and to the receiver antenna input. The phasing control is a potentiometer (R1). The value of this pot is usually 10 to 100 kohm, with 25 kohm being a commonly seen value. Switch S1 is used to take the sense antenna out of the circuit. The reason for this switch is that the nulls of the loop or loopstick are typically a lot deeper than the null on the cardioid pattern. The null is first located with the switch open. When the switch is closed you can tell by the receiver S-meter whether or not the correct null was used. If not, then reverse the direction of the antenna and try again.

SWITCHED PATTERN RDF ANTENNAS

Suppose we have a unidirectional pattern such as the cardioid shown in Figure 11.18. If we can rapidly switch the pattern back and forth between two directions that are 180 degrees apart, then we can not only discern direction, but tell whether an off-axis signal is left or right. This feature is especially useful for mobile and portable direction finding.

In Figure 11.18 we have three different positions for a signal source. When the signal source is at point A, it will affect the pattern to the left more than the pattern to the right, so the meter will read "LEFT." If the signal source is at point B, on the other hand, the signal affects both pattern positions equally, so the meter reads "ZERO." Finally, if the signal arrives from point C, it affects the right hand pattern more than the left hand pattern, so the meter reads "RIGHT."

Figure 11.19 shows how such a system can be constructed. This system has been used by amateur radio operators using "rubber ducky" VHF antennas and a single receiver, where it is commonly called the "Double-Ducky Direction Finder" (DDDF). The antennas are spaced from 0.25λ to 1λ apart over a good ground plane (such as the roof of a car or truck). If no ground plane exists, then a sheet-metal ground plane should be provided.

In Figure 11.19 we see the antennas are fed from a common transmission to the re-

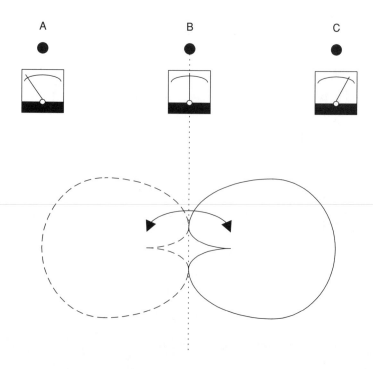

Fig. 11.18
Switched pattern RDF patterns.

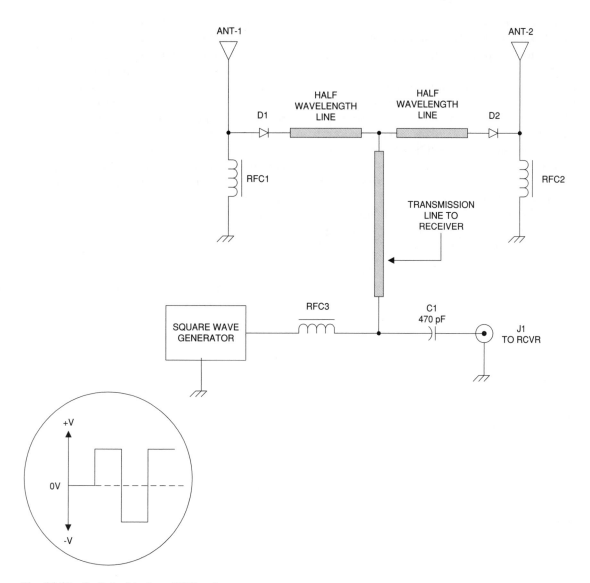

Fig. 11.19 *Switched pattern RDF system.*

ceiver. In order to keep them electrically the same distance apart, a pair of identical half-wavelength sections of transmission line are used to couple to the antennas.

Switching is accomplished by using a bipolar square wave and PIN diodes. The bipolar square wave (see inset to Figure 11.19) has a positive peak voltage and a negative peak voltage on opposite halves of the cycle. The PIN diodes (D1 and D2) are connected in opposite polarity to each other.

Diode D1 will conduct on negative excursions of the square wave, while D2 conducts on positive excursions. The antenna connected to the conducting diode is the one that is connected to the receiver, while the other one is parasitic. The active antenna therefore switches back and forth between ANT-1 and ANT-2.

This antenna is coupled to the receiver through a small value capacitor (C1) so that the square wave does not enter the receiver. This allows us to use the transmission line for both the RF signal and the switching signal. An RF choke (RFC3) is used to keep RF from the antenna from entering the square wave generator.

The DDDF antenna produces a phase modulation of the incoming signal that has the same frequency as the square wave. This signal can be heard in the receiver output. When the signal's direction of arrival is perpendicular to the line between the two antennas, the phase difference is zero, so the audio tone disappears.

The pattern of the DDDF antenna is bidirectional, so there is the same ambiguity problem as exists with loop antennas. The ambiguity can be resolved by either of two methods. First, a reflector can be placed $\lambda/4$ behind the antennas. This is attractive, but it tends to distort the antenna pattern a little bit. The other method is to rotate the antenna through 90 degrees (or walk an L-shaped path).

One of the uses of this type of system is that it can be fitted to a VHF scanner receiver for portable RDF efforts at locating EMI emitters.

Chapter 12

EMI to Television, Cable TV, and VCR Equipment

For many people, EMI to video and television equipment is all there is to the subject. Video and TVI is easily seen or heard and is terribly irritating to a large number of people. In this chapter we will take a look at video/TV interference and what can be done about it. Considered will be antenna connected television receivers, cable television receivers, and VCR equipment.

THE BASIC TELEVISION RECEIVER

Before we can deal with television interference, we must first deal with the standard television system. In the United States and Canada we use NTSC television, while foreign countries use SECAM or PAL systems. Some wags tell us the U.S./Canada standard means "never twice same color," but in reality it means "National Television Steering Committee." Figure 12.1 shows the basic NTSC color TV signal. The FCC has allocated space based on this model for transmission of TV.

The NTSC color TV standard requires a video carrier 1.25 MHz from the lower end of the channel. Spaced 3.579 MHz from the video carrier is a color burst carrier, which is not transmitted (it is suppressed in the transmitter). Spaced 4.5 MHz from the video carrier is an audio carrier. This latter is frequency modulated (FM) with a 25-kHz deviation. The video carrier is vestigial sideband (a species of AM in which one sideband is produced in full and one sideband is produced as a vestigial sideband). The chroma information is the color signal and is phase modulated onto the signal.

A typical television receiver is shown in block diagram form in Figure 12.2. This receiver is a superheterodyne, and so will have a front end inside the tuner, which is downconverted to an IF frequency. An AM envelope detector circuit called a video detector is used to demodulate the IF signal. It is the video detector that separates the video, color, and sound information.

The video amplifier section takes the composite black-and-white video and amplifies

135

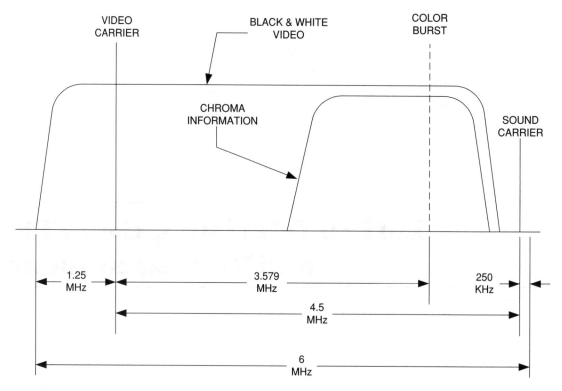

Fig. 12.1 *Standard U.S. television channel.*

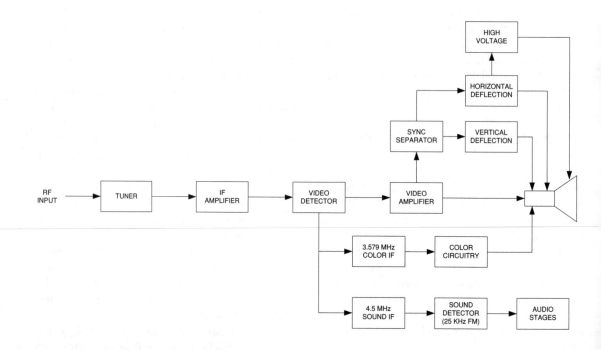

Fig. 12.2 *Standard U.S. television set.*

it to a point where it can drive the cathode ray tube (CRT). As part of the video amplifier or detector (shown as part of the amplifier here) is a circuit called a sync separator. This circuit will separate out the horizontal and vertical deflection pulses.

The deflection amplifiers control the vertical and horizontal aspects of the CRT. The vertical deflection is at 59.94 Hz, and the horizontal deflection is 15,734 Hz (these were changed from 60 Hz and 15,750 Hz to accommodate color). The high-voltage DC for the second anode of the CRT is derived from the horizontal amplifier circuit.

The color information is passed through a 3.579-MHz IF amplifier from the video detector. From there, it is processed and displayed on the CRT in the form of (usually cathode driven) red, blue, and green signals.

The sound information is passed through a 4.5-MHz IF amplifier and a 25-kHz FM detector to form the sound signal. From there, the audio signal is built up in a chain of amplifiers and output to a loudspeaker.

The front end of the television set is the tuner. It consists of a mixer circuit, a local oscillator, and (usually) an RF amplifier circuit. The front end is fed with the RF transmission line from the antenna. Either 300-ohm twin-lead or 75-ohm coaxial cable may be used. The latter case, 75-ohm coaxial cable, predominates today, but one can still find older television receivers and low-cost modern television receivers that use 300-ohm twin-lead.

FUNDAMENTAL OVERLOAD

Fundamental overload of a television receiver occurs when there is a strong signal present at the input of the receiver. The television receiver responds to the fundamental of the offending transmitter. It takes 25 to 30 dB of signal-to-noise ratio (SNR) to make a decent TV picture, and if there is less than that magic number, interference can result. The symptoms of fundamental overload include a herringbone pattern, or even a blanked-out picture.

HARMONIC OVERLOAD

Although fundamental overload is the most common form of EMI to TV receivers, harmonics from the transmitter may be second (especially with HF or low-band VHF transmitters). Harmonics are integer multiples of the fundamental frequency of the transmitter, and they can reach considerable heights in transmitters that are misadjusted. The FCC requires transmitter owners to keep the harmonic content of their emissions down −40 dBc (decibels below carrier) to −60 dBc, depending on service and frequency.

Harmonic overload may occur because of misadjustment of the transmitter. It may also be generated in the television receiver itself if there is a PN junction in the circuit that can be biased into nonlinearity by the transmitter signal. It may also be generated by dirty connections, corrosion, or even rusty bolts.

AUDIO RECTIFICATION

Audio rectification is heard rather than seen. It is audio from the offending transmitter's modulation being picked up, envelope detected, and amplified by the TV's audio amplifier chain. Interestingly enough, audio rectification occurs more often in cases where the antenna of the offending transmitter is vertically polarized. A good indication that interference is audio rectification is that it is audible regardless of the setting of the volume control.

IMD INTERFERENCE

A television receiver is an amplifier and demodulator of a radio signal, so it is susceptible to intermodulation distortion (IMD) interference. IMD interference occurs when two or more signals are present in the input of the receiver. They produce additional frequencies according to $mF_1 \pm nF_2$, where F_1 and F_2 are the frequencies and m and n are either integers or zero. That means a lot of frequencies! If two or more frequencies happen to follow this rule and produce a third frequency that is on a television frequency, then IMD interference will result.

IF INTERFERENCE

IF interference exists because of the IF amplifier chain in the television set. It is easily identified because it affects all of the TV channels. Furthermore, it produces a fine herringbone pattern on the screen that rotates with the setting of the fine-tuning control. No other form of interference does this trick (although with other forms of interference the intensity of the interference does change with the fine-tuning).

DIRECT PICKUP

This form of interference, which requires transmitters of several kilowatts to accomplish, is picked up directly by the television receiver. It is generally due to internal wiring or printed circuit boards. To test for this type of interference, terminate the antenna in its characteristic impedance and note whether or not the signal level decreases. If it does, then the signal is arriving via the antenna terminals, and it is not a case of direct pickup.

COMMON-MODE VS DIFFERENTIAL-MODE SIGNALS

Very little has been written about the difference between common-mode and differential-mode interference. The difference is simple: differential-mode interference arrives entirely inside the feedline, while common-mode interference does not. The common-mode signal may arrive on the transmission line or on the AC line cord. The treatment of each form of interference depends on whether it is common or differential mode. It may be true that both forms of fix are required in a given case.

COMMON-MODE FILTERS

Common-mode filters come in two varieties: transmission line and AC line cord. Both are used to suppress EMI in TV systems. The transmission line variety is shown in Figure 12.3. It consists of a toroid core wrapped with the television antenna transmission line. This common-mode choke should be installed as close to the TV set as possible. To make an AC line cord type of common-mode choke, wrap the toroid with the AC line cord as close as possible to the TV set.

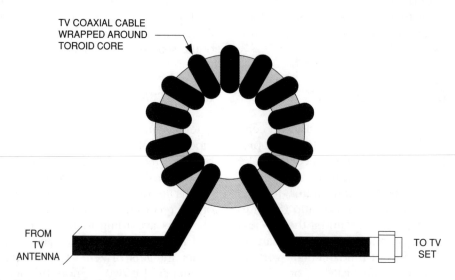

TV COAXIAL CABLE
WRAPPED AROUND
TOROID CORE

FROM
TV
ANTENNA

TO TV
SET

Fig. 12.3   Common-mode choke.

FILTERING

The television can be filtered at the antenna terminals or the AC line cord. For further information on the latter case, see Chapter 7. In this chapter we will discuss the type of filtering that one places on the transmission line.

Figure 12.4 shows the proper installation of the filter at the TV receiver. Figure 12.4A shows a 300-ohm system, while Figure 12.4B shows a 75-ohm system. The key in both cases is to make the transmission line between the television and filter *as short as possible.* In fact, it would be preferable to install the filter right at the antenna terminals of the tuner inside the TV set, but that is not usually possible unless you are a television repairman.

The type of filter to install depends on the nature of the transmitter causing interference to the particular TV set. In cases where the interference is being caused by an HF band transmitter (amateur, CB, or commercial), use a high-pass filter at the television set. It should have a cutoff frequency below TV channel 2. For VHF or UHF interference, use a high-pass, low-pass, bandpass, or bandstop (i.e., notch) filter, depending on the situation. If the offending transmitter affects more than one channel, it may be necessary to use a bandstop filter centered on the transmitter's frequency at the TV receiver.

Figure 12.5 shows a type of filter that works well for both common-mode and differential-mode signals in coaxial cable type systems. It consists of a 300-ohm twin-lead type of high-pass filter sandwiched between

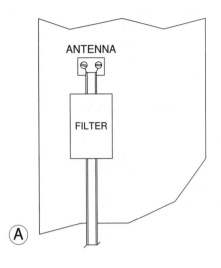

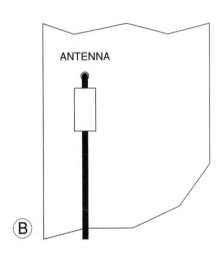

Fig. 12.4 *Correct placement of filters: (A) 300-ohm twin-lead version; (B) coaxial version.*

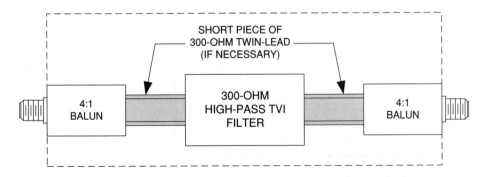

Fig. 12.5 *Balanced filter circuit.*

two 4:1 BALUN transformers. The 300-ohm twin-lead between the transformers and the filter should be as short as possible, or of zero length if possible.

STUBS FOR EMI ELIMINATION

Stubs can be used for EMI suppression on TV receivers. The usual quarter- and half-wavelength stubs are used, but Figure 12.6 shows a one-eighth wavelength stub. It is used to eliminate one particular frequency, and so it

functions as a notch filter. The frequency of the notch filter should be the frequency of the offending transmitter. The stub has a capacitor (C1) at the end, and this capacitor is used to tune the stub to the right frequency.

FARADAY SHIELDED COAXIAL CABLE

Figure 12.7 shows the use of a Faraday shielded coaxial cable for suppression of EMI. This forms a species of common-mode choke. To make the choke, take a short length of coaxial

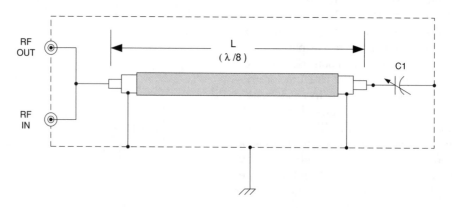

Fig. 12.6   *One-eighth wavelength stub.*

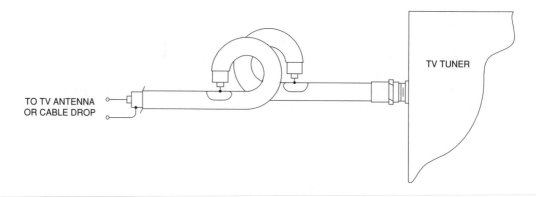

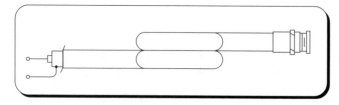

Fig. 12.7   *Faraday shielded TV input.*

cable with a connector on one end. Wrap the end into a loop of less than 6 inches in diameter. Scrap away a contact point on the outer insulator and connect the inner connector to it. Next, do the same thing to the cable from the antenna or cable TV customer drop. Lay the two cables over each other in the manner shown on the inset to Figure 12.7 and secure with tape or cable ties.

CABLE TELEVISION SYSTEMS

The cable television system in the United States is a closed system used to carry multiple channels to homes, businesses, and other subscribers. It should be very clean, but often isn't. The fact that they share certain channels with other services (amateurs, navigation, and communications) makes the system ripe for interference.

The basic cable system is shown in Figure 12.8. It can be coaxial cable or fiber optic in nature, of which the fiber optic is by far most free of EMI problems. It is also the rarity. Although in some areas fiber has replaced coaxial cable, most cable systems are made of coaxial cable.

The head end of the cable system takes programming from various sources (off-the-air channels, satellite, local origin), and passes them to various trunk lines. Only one trunk is shown for clarity. The signal on the trunk is relatively low level, but is boosted by trunk amplifiers before being distributed further. From each trunk amplifier there is a bridger

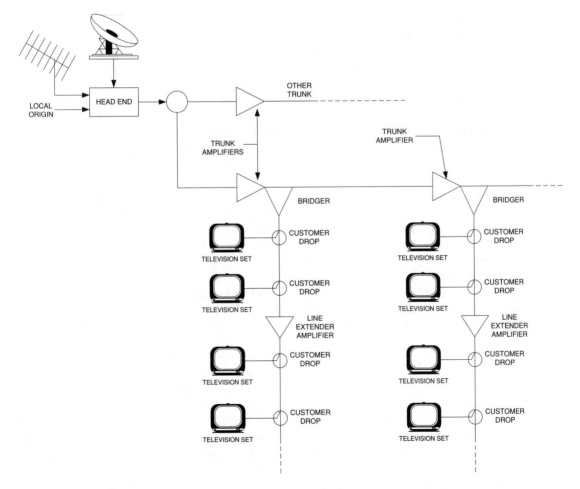

Fig. 12.8   *Cable TV system.*

network that distributes signal to the various subscribers. The subscriber tap into the system is called a drop. A line extender amplifier is used to boost the signal lost in the coaxial cable and various drops to a level that is useful in the system.

The quality of the signal is determined by two factors: noise and distortion. The noise is mostly that of the amplifiers. It degrades picture quality. Distortion can be harmonic or IMD, but is mostly IMD. The length of each distribution leg is limited principally by the distortion characteristics of the amplifiers.

TWO-WAY CATV

It is possible for the cable TV provider to communicate two ways on their systems. The 5- to 40-MHz region is used to provide upstream communications between the subscriber and the cable TV provider, and the institutional net (or I-net) in the sub-split system (in the mid-split system 5 to 112 MHz is used, and in the high split 5 to 174 MHz is used).

CHANNELIZATION

There are three generally accepted channelization schemes used in cable TV systems: standard, harmonically related carrier (HRC), and incrementally related carrier (IRC). These are defined in standard EIA-542. Of these, the standard is by far the most common. It uses TV channels assigned from 54 MHz in $6n + 1.25$ MHz schemes, up to 1,002 MHz. The HRC method uses TV channels locked to a 6.000 MHz comb. The IRC uses the basic $6n + 1.25$ MHz carriers locked to an internal comb.

LEAKAGE

Cable TV is supposed to be a closed system. Cable TV leakage occurs, however. It is called ingress when an outside source interferes with the cable TV, and egress when the RF energy gets out. Frequencies from 5 to 750 MHz are used, including some assigned

Table 12.1   CATV Leakage Limits

Frequency (MHz)	Distance (ft)	Field Strength (mV/m)
5–54	100	15
54–216	10	20
216–750	100	15

to other services. Leakage limits are related to distance in Table 12.1.

A field strength of 15 mV/m at a distance of 100 feet from the cable TV is a strong signal. But the regulations also maintain that the cable TV system must not cause harmful interference to other services. The definition of "harmful" is not given, however.

RESPONSIBILITY

The responsibility for EMI to and from cable TV systems is split among the cable TV system, the transmitter owner, and the owner of the afflicted television receiver. The transmitter owner's responsibility is limited by the terms of his or her license. If the transmitter is operated according to the terms of that license, including the emission of spurious signals, then there is little one can do to the transmitter owner. That leaves the cable TV system and the subscriber. However, cooperation of the offending transmitter owner is needed to troubleshoot the interference problem.

FINDING LEAKS

The problems with cable TV ingress and egress start with the effects of weathering on the system, as well as intentional acts of the customer. To protect against the former the FCC requires cable TV operators to continuously patrol the system to find leaks. Finding the leaks is an exercise in radio direction finding. The technician will drive along the line while monitoring the leakage levels on a sensitive receiver. The mobile unit will locate the leak within about 80 to 100 feet, after which hand-held receivers of lower sensitivity are required.

WHAT TO DO WHEN THE INTERFERENCE IS AT THE SUBSCRIBER END

What to do when the problem is at the subscriber end is a difficult problem. Some cable TV companies are willing to troubleshoot the situation to keep the subscriber happy, while others are not. Modifications to the television set should be done by competent personnel, but beyond that it is possible to troubleshoot the system.

First, diagnose the problem. Is the set or the converter really involved? The interference may be experienced by only one subscriber. In that case, assume that the subscriber's system is at fault (multiple faults along a line do not necessarily mean that the system is at fault, but do point to a system problem).

The basic fix for most problems is a common-mode choke. The common-mode choke can be made using the cable and a toroidal core per instructions given in Figure 12.3. It consists of a toroid core wrapped with coaxial cable. Figure 12.9 shows the placement of the common-mode choke in the system. This does not preclude placing a common-mode choke and a differential-mode high-pass filter at the television receiver's antenna input terminal.

There are two paths for interference to occur: power line conduction and direct radiation. The power line conduction scheme can be eliminated at the transmitter by placing either a common-mode choke or a differ-

ential filter in the power line. In other cases, however, the common-mode choke should be placed in the AC line cord of the converter or susceptible TV receiver. The common-mode choke is made by winding the AC line cord around the toroid core.

WHAT TO DO WHEN THE CUSTOMER IS AT FAULT

There are two cases where the customer might be at fault: (1) the existence of a susceptible TV and (2) the use of multiple sources. The case of the susceptible TV is usually found by disconnecting the converter from the antenna terminals and replacing it with a carbon resistor of either 300 ohms or 75 ohms (depending on the impedance of the set). If the interference persists, then the set is responsible. In that case, common-mode chokes on the antenna terminal and AC line cord, and differential-mode filtering at the antenna terminals, may help.

The use of multiple sources for their TV signals may be the problem. Look for connections that should not be present. If the connections are made through an A/B switch, then disconnect the A/B switch and connect the converter directly. If the interference disappears, then you have solved the problem. If not, keep trying!

VCRS

The VCR is a special device that is placed in the TV system to record signals off the air or play back tapes. The two systems in the

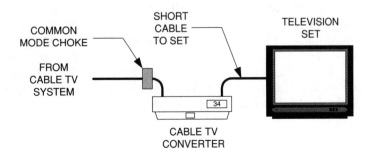

Fig. 12.9
Common-mode choke in cable TV system.

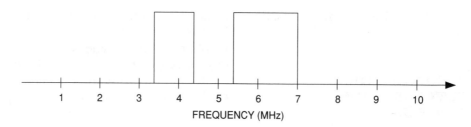

Fig. 12.10   *Luminance signal in straight and Super VHS VCRs.*

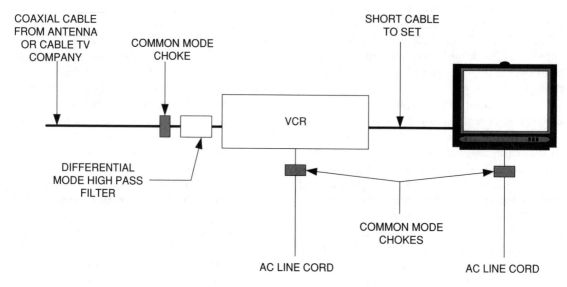

Fig. 12.11   *Placement of chokes in VCR system.*

United States and most of the world are VHS and Super VHS. They present certain problems to the troubleshooter.

The easiest way to record the 6-MHz bandwidth onto tape is to use frequency modulation with the baseband video signal. Of course, one can tape-record it directly, but that would be prohibitively expensive on consumer-grade equipment. The luminance (brightness) signal is recorded from 3.4 to 4.4 MHz in VHF, and 5.4 to 7.0 MHz in Super VHS (see Figure 12.10). Some VCRs will record the audio portion separately from the video portion.

Figure 12.11 shows the approach for attacking the EMI problem to VCRs. Common-mode chokes should be placed in the AC line cords of the television and the VCR, as appropriate. A common-mode choke should also be placed at the antenna input connector of the VCR. If differential-mode signals exist, then use a differential-mode high-pass filter (or band-reject filter in the case of VHF transmitters).

It is uncommon to find this, but be aware that a VCR is a transmitter, and if the output of the VCR is connected to the antenna, it will radiate on the channel being viewed. This is usually channel 3 or 4. This means that the neighborhood will view the same pictures as the original viewer! The antenna should go through the VCR even if the user does not intend to record off the air.

Chapter 13

EMI to Consumer Electronics

The phrase "consumer electronics" can mean a number of different things, but most of them are audio. This chapter is about interference to audio equipment, whether stereo high-fidelity equipment or the intercom used to announce front-door visitors.

ROLES AND RESPONSIBILITIES

The FCC does not regulate what happens when a radio transmitter interferes with audio equipment. The issue is seen as being between the parties involved and not a matter for FCC concern. The reason is that the interference is due to the affected audio equipment not performing according to the following:

1. It must respond to signals that are intended for it

2. It must reject unwanted signals

It is in the second of these parameters that the equipment fails. It performs the first of these quite well, but fails miserably in the second. The FCC only regulates radio interference to other services' receivers, so it will not usually get involved in disputes involving audio equipment. The only exception to this rule is where the interference is to the receiver in the stereo system, and that interference is not due to audio rectification.

The FCC does, however, recognize that there will be a lot of calls to their field engineering offices and headquarters regarding audio interference. As a result, they have prepared a booklet on the subject. It can be accessed on the Web at http://www.fcc.gov.

The transmitter owner has no legal responsibility to not interfere with audio equipment. There is, however, a moral obligation to cooperate with you as you troubleshoot the problem (if only to ensure that transmissions will exist during the time when you are present).

Audio interference is probably the only form of interference where the equipment owner is totally responsible for the fix. The equipment does not meet its requirements, but the owner only sees the fact that the interference occurs only when the transmitter is on the air. That may make for some interesting conversations! After the equipment owner acknowledges that he is responsible for the fix, then it's possible to do the fix . . . and it's quite possible that the fix will be successful.

THE SOURCE OF THE PROBLEM

The source of the problem of audio rectification is (a) that there are PN junctions in the circuit, or (b) that the amplifier may be driven into nonlinearity by the RF signal, detecting the AM signal as if it were a diode. In either case, nonlinear behavior will afflict the otherwise linear amplifier, and that causes problems. Audio rectification is not the transmitter operator's fault, but rather is the fault of the equipment being interfered with. The equipment is not able to meet criterion number 2 above: It must reject unwanted signals.

TYPICAL AUDIO SYSTEM

Figure 13.1 shows the block diagram for a standard audio system such as might be encountered in a residence. There will be at least two channels on a stereo system. The main amplifier is a power amplifier capable of generating 0.5 to several hundred watts. Control of the source and bass/treble profile is provided by the preamplifier stage. It con-

tains the bass, treble, balance, and volume controls. It takes the low input from the sources and boosts it to the level required to drive the main power amplifier (100 to 500 mV). The sources can be few or many depending on the system. Typical devices hooked to the input of the preamplifier include cassette tape, reel-to-reel tape, TV set, CD-ROM, DVD, radio receiver, and (less often now) a turntable for phonograph records.

In a typical system the preamplifiers and main amplifiers are part of the same body, along with the radio receiver. This is usually called a receiver, but has the preamplifier and main amplifier inside. Other units have separate preamplifier and main amplifier portions.

It is the interconnecting wires that form the basis of EMI problems in stereo systems. The input wires are shielded, but often not to RF standards. The speaker leads are usually not shielded and are often far too long. That makes them act like an antenna any frequency where they are a substantial fraction of a wavelength. Added to these problems is that the speaker leads are hooked to a feedback network inside the amplifier. Figure 13.2 shows this system. Transistor Q1 is a preamplifier or driver transistor, and Q2/Q3 is the output stage of the amplifier. The feedback network is essential to keeping down the distortion and noise in the system, but it tends to be a high-pass filter.

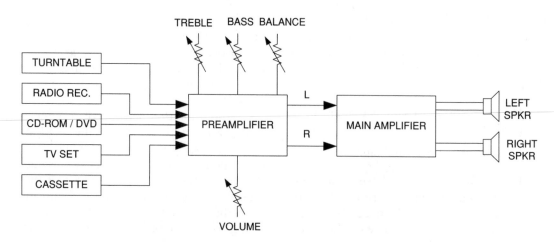

Fig. 13.1 *Typical stereo system.*

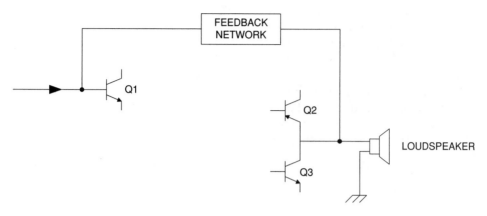

Fig. 13.2   *Feedback network is basically a high-pass filter.*

PATHWAYS FOR TROUBLE

There are three pathways for EMI to enter the stereo system: radiated, conducted, and magnetic induction. Magnetic induction is not usually a problem unless there is a source of interference very near the stereo, and only then because of defects in the stereo.

The conducted signal arrives by wire and may be either common mode or differential mode (although usually common mode). It is rare for conducted interference to be the only form of interference in a system. There almost always is a radiated component.

The radiated signal is the most common form of interference to audio systems. The nearby transmitter will radiate a signal to the wires making up the interconnections of the audio system, causing the interference. Wires make good antennas! These interconnecting wires will pick up the signal and, if the system is susceptible, will cause interference. The quick fix is to shorten the wires that cause the interference. Shorter wires will pick up less signal, whether it is common mode or differential mode. Don't just stow the excess in a pile behind the amplifier; at least roll it up into a coil if you can't shorten it. A common-mode choke, placed right where the leads go into the amplifier, is a viable alternative.

There is a form of radiation pickup that is a special case. The stereo or other equipment will sometimes pick up radiation directly on its printed and nonprinted wiring. This is called direct radiation pickup. It is difficult to cure because no common-mode or differential-mode filter in line with a lead or connection will suffice. The only thing that will cure this form of radiation pickup is shielding of the affected circuits.

TROUBLESHOOTING

The key to troubleshooting this problem is to note whether the volume control has an effect on the interference signal. If the volume control has no effect on the interfering signal, then the problem is likely to be audio rectification in the main amplifier (or, at least, in the preamplifier beyond the volume control). Alternatively, the interference could be due to pickup on the AC line cord or a ground loop.

If the volume control does have an effect on the interfering signal, then the problem is before the volume control. It could be common-mode or differential-mode interference prior to the volume control, or it may be direct pickup by the high-gain preamplifier stages.

Troubleshooting begins in the case of interference that is affected by the volume control by disconnecting all of the leads coming from the sources. Next, reconnect each in its turn while noting the interference level. Disconnect each source between trials so that an accurate picture is formed (it is rare to find only one source of interference).

If the interference exists when the set is turned off, then suspect that the speaker leads are picking up and audio rectifying the RF signal. The PN junctions in the transistors are audio rectifying the signal, and they are coming through unamplified.

CURES

The cure that you will use depends upon whether the interference is common mode or differential mode. A variety of common-mode and differential-mode filters should be kept on hand. The common-mode choke should be used in any input that causes a problem with EMI, as well as the speaker leads.

The common-mode choke consists of a ferrite core wrapped with the lead in question. The type of core doesn't matter much, except that the rod will radiate into surrounding circuitry whereas the toroid will not. Use FT-240 or another grade of ferrite that will present a high impedance to RF signals and a low impedance to audio.

Ferrite beads are an option for those cases where you are able to get into the equipment and make modifications. They act like an RF choke in series with any wire they are placed over. Figure 13.3 shows the use of a ferrite bead at a transistor should a transistor be available for such treatment. Place the ferrite bead over the base terminal of the transistor in most cases. Mix 43, 73, and 75

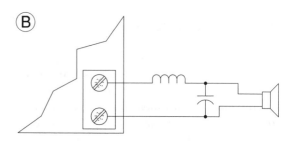

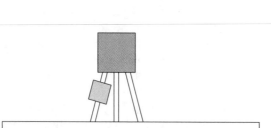

Fig. 13.4   *(A) Don't bypass speaker with a capacitor; (B) use an L-C network.*

are most frequently used for EMI problems in circuits.

The idea of bypassing the speaker leads of solid-state amplifiers with a simple capacitor (Figure 13.4A) should be avoided (the same advice does not hold true for vacuum-tube amplifiers). The reason is that they interfere with the operation of the feedback network and may cause the amplifier to break into full-power oscillation. A better solution is to use either a common-mode choke or an L-section filter in which the lead item is an inductor (Figure 13.4B).

Figure 13.5 shows two circuits for use inside the equipment. In each case (Figures 13.5A and 13.5B) the audio source is one of the devices mentioned earlier. In Figure 13.5A we see an R-C filter based on a 150-ohm resistor and a pair of capacitors. Keep the value of the capacitors low, starting at 100 pF and working up. As it is, the circuit

Fig. 13.3   *Ferrite bead on transistor leg.*

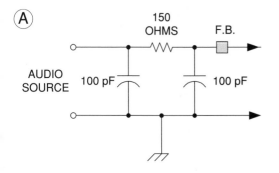

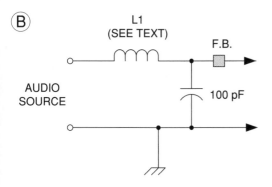

Fig. 13.5 *Typical filter circuits used in input circuits.*

has a resonance about 10 MHz, above which the response falls off. The filter in Figure

13.5B shows a version using a coil and a capacitor. Again, the capacitor should be as small as possible. The value of coil L1 is determined by the frequency of the interference source, but in no event should be higher than about 2.5 mH.

SHIELDING

Shielding works well in cases of direct radiation pickup, but it is difficult to effect on audio equipment. It is the rare person, indeed, who will allow the audio equipment to be operated inside of a shielded cabinet! The alternative is to shield the affected equipment inside the cabinet. On nonmetallic cabinets, this can be done most effectively with a conductive spray.

GROUNDING

The existence of ground loops is to be avoided where possible. The use of a star ground is mandatory in these cases. Ground all items to the main amplifier or preamplifier, as appropriate.

Chapter 14

EMI from Computers

Computers are a fact of life in the modern workplace or home. They can be purchased for around $1,000, so most people can afford them. As new software is developed, we find that the modern personal computer is a very powerful adjunct to modern living. Unfortunately, computers are digital devices and as such produce EMI. The degree of EMI is indicated by the fact that current computer speeds sound like radar frequencies (450 MHz)! This means that there are RF signals on the inside of computers that must be shielded or filtered. Much of the filtering or shielding will be the same as for other products and so will not be discussed again here. What we will cover in this chapter is special features of computers.

THE LAW

In the United States the effects of EMI from digital devices are regulated under 15 CFR (Code of Federal Regulations). There are two classes of devices (Class A and Class B). Class A regulations are less stringent than Class B because they are intended to cover devices used in commercial environments (Class-B devices are residential devices). The regulations for Class-B residential devices are at least 10 dB more stringent frequency-for-frequency than those for Class-A devices.

In Europe, the CE Mark indicates that the device has passed certain EMI standards. As I understand the European situation, it is illegal to sell devices that are not CE Marked in Europe.

The FCC maintains a Public Access Link on the World Wide Web that deals with many issues, among them computers and EMI. The FCC link is http://www.fcc.gov; navigate to the Public Access Link.

THE PROBLEMS WITH COMPUTERS

Several problems are associated with computers, and these result in either radiated or conducted EMI problems. One of the principal problems derives from the fact that computers are digital devices and as such have clock pulses associated with them. These clock pulses are square waves (Figure 14.1A) and as such possess harmonics (Figure 14.1B). To further complicate matters, the clock is divided in frequency to lower values by the CPU

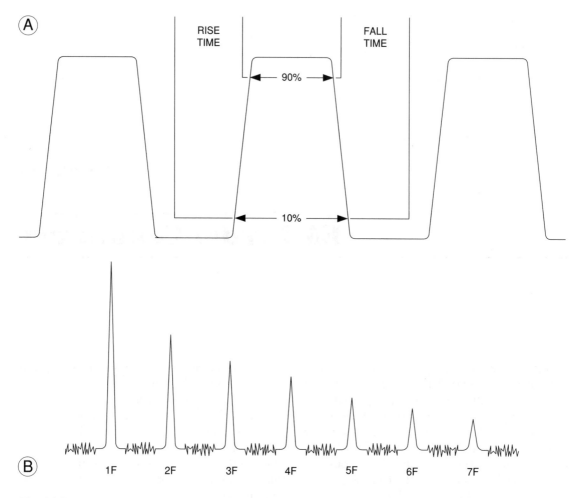

Fig. 14.1   *(A) Rise time and fall time defined; (B) spectrum of harmonics.*

chip inside the computer. As a result, there are numerous frequencies that will be afflicted by EMI from harmonics.

The problem with square waves is in the rise time of the waveform (see Figure 14.1A again). The rise time and fall time are assumed to be symmetrical for this discussion. The rise time is defined as the time required for the wave form to rise from 10 percent to 90 percent of the total amplitude (the fall time is similarly defined). The spectral energy falls off at a rate of 10 dB/decade at frequencies determined by:

$$F_{CO} = \frac{0.318}{T_r}$$

where:

F_{CO} is the fundamental frequency
T_r is the rise time
. . . and 20 dB/decade at frequencies above this point

Clearly, the square wave is a major contributor to EMI problems. The harmonics and subharmonics of the clock system are capable of interfering with frequencies that are considerably higher and lower than the fundamental. The problem in the medium-wave and high-frequency shortwave bands is alleviated with high-frequency clocks, but the problem in the VHF/UHF/microwave spectrum is enhanced.

CABINETRY

Figure 14.2 shows the typical computer case with the key points of interest. The motherboard and the plug-in boards contain the EMI-producing components, for the most part. Key to an low-EMI environment is:

1. A cabinet with adequate "finger" grounds on the removable cover

2. Metal (not plastic) covers for the bays that are not used

3. Keeping the unused slots for plug-in cards plugged with a metal strip

The switching DC power supply is a source of interference in some cases, but as this part is usually shielded it is usually a non-player. It will radiate if the cabinet is not well shielded, but at a frequency about 50 kHz.

The power supply is nothing compared with the monitor as a source of EMI from computer systems. In fact, the monitor deflection system is a primary cause of EMI in

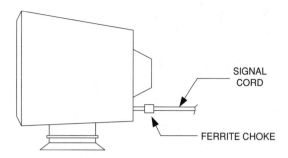

Fig. 14.3   Ferrite common-mode choke on monitor.

the lower frequency regions. There are three things that can be done to ensure a quiet computer monitor:

1. Buy a monitor that has internal shielding around the deflection circuits

2. Buy a monitor that has a screened viewing surface

3. Buy a monitor that uses ferrite choke (Figure 14.3) in the signal cord to the computer

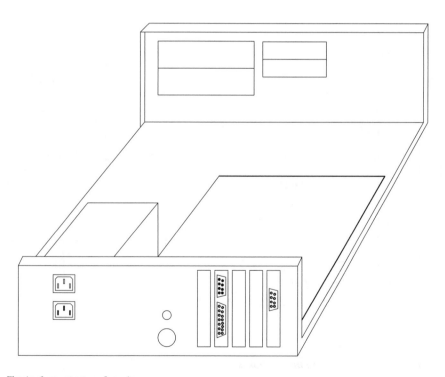

Fig. 14.2   Typical computer chassis.

TROUBLESHOOTING COMPUTER EMI

If the computer is known to be a problem, there are some things that can be done to eliminate (or at least mitigate) the problem. If the source is suspected (or known) to be a specific computer, then turn off the computer to see if the noise disappears. In fact, you should turn off the monitor and computer separately to see which is at fault.

Once the source of EMI is determined to be a specific computer, it is necessary to troubleshoot the source of the problem within the computer. For this test, the computer should be operating in the normal manner. Disconnect the lines from the computer, including the keyboard and mouse, one by one to see if there is a significant decrease or increase in the noise level. The cable (if any) that causes the problem should be shielded.

If the noise persists after the cables are disconnected, it must be direct chassis radia-

tion. If the case is made of metal, you should make sure that all screws are tight and in place. Small holes are generally not a problem, but long slits are a problem. If you find a slit or hole, cover it with conductive foil. If the case is plastic, you will need to shield the inside of the case (conductive paint will do this job).

THE ROLE OF FERRITE CHOKES

The role of ferrite cores in EMI reduction cannot be overemphasized. These cores are either clamp-on or in-line, but they share the capability of acting like RF chokes in the line. Figure 14.4 shows the typical computer system in which ferrite cores are used to eliminate EMI. The mouse should be arranged such that the ferrite core is at the computer end, while the other devices may require two cores in order to be effective. The problem revolves around whether or

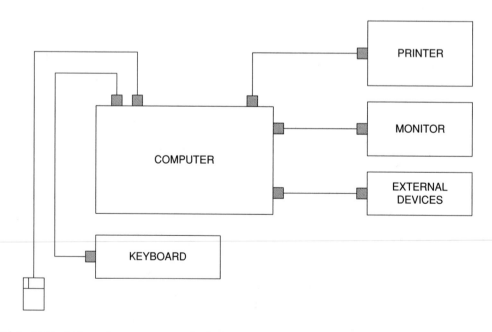

Fig. 14.4 *Typical places for common-mode chokes.*

not the noise originates inside the device or outside.

GROUND LOOPS

Normally computers are grounded through their electrical power cords. This is sufficient for most cases. But in cases of EMI it might not be sufficient. In those cases try to bond the equipment together in a single-point grounding system. The National Electric Code will be of use in this matter, but keep in mind that a good ground at HF or less is not the same as a good ground at VHF/UHF frequencies. At the higher frequencies, the length needed to make a ground connection is a significant fraction of one wavelength . . . or more.

EMI TO COMPUTERS

If RF energy can get out of computers, then it is possible for RF energy to get in . . . that's elementary. Key a walkie-talkie near a computer and its operation may be seriously disrupted. The principal paths of the RF into the computer are the same as those for getting out, and the fixes are the same.

Two particular routes of EMI are the modem connector and the AC line cord. In either case the fix is an appropriate filter placed as close to the computer cabinet as possible. In the case of the AC line cord filter, make sure that the filter can handle the power requirements of the computer or you will create an overload condition.

Chapter 15

Mystery EMI: Rusty Downspouts and All That

This chapter deals with some of the EMI oddities that I've seen or heard about over the years. One of the things that comes along with radio communications is electromagnetic interference (EMI), and the necessity for electronic products (communications and otherwise) to possess electromagnetic compatibility (EMC). It's not always so easy. The problems of EMI/EMC afflict all communications: commercial, governmental, citizens band, and amateur radio. It's one of the things these services have in common.

Over the years I've serviced a lot of different problems of EMI/EMC, including residential, business, industrial, and mobile. Some of them are pretty funny.

MOODIE AND THE CROWN VICKIE

During a much earlier period of my life I worked installing both citizens band and landmobile two-way radios, as well as ordinary automobile radios. One of the main jobs for an installer of mobile electronic gear is to locate and suppress interference sources.

And vehicles abound in such sources! The ignition and the charging system are prime culprits, but also causing problems are things like the gas-gauge sending unit, power windows, and almost anything else electrical. Today, we have a number of digital processors and computers on board as well as the traditional noise sources.

Even if your field of interest is limited to eliminating mobile ignition system noise, the task can be daunting. I've seen cases where an ungrounded hood caused massive noise problems. And the fiberglass hoods found on some cars are absolutely evil if the bonding comes loose! In other cases, noise is induced on the DC power lines that pass through the firewall from the engine compartment to the passenger compartment, where it is reradiated and picked up by the electronic equipment. In some cases, an ungrounded engine exhaust pipe will reradiate noise as effectively as an antenna.

All of those things are routinely found. But some are not so routine. Once (if you will permit me a nostalgic regression), I was working installing CB sets at the dawn of the CB era (late 1950s and early 1960s). The vehicle was a 1956 Ford Crown Victoria sedan (which was *not* an antique car in those days). I tried everything in the technician's bag of tricks, and the CB kept clicking with ignition noise all across the band. The master technician, a rough and ready fellow named Moodie, came down to the garage, determined to "show that Carr kid how it's done." He inspected my work and could find no fault. He tried a few things himself, and after two hours was still unsuccessful. At that point, weary from lack of success (not to mention a two-hour chewing out by Moodie), I leaned my elbow against the chrome roofline of the Crown Vicky. The noise stopped! One of the features that distinguished the '56 Crown Vicky from less costly models was a 9-foot-long curved chrome decoration strip around the front of the roof line, continuing on to the two sides of the vehicle. Get the point? Nine feet is quarter-wavelength at the 11-meter (27-MHz) citizens band, so even minute amounts of radiation would find a resonant situation and reradiate right into the antenna! Cleaning and resetting the clips and screws that held the chrome strip fast solved the problem . . . and we were able to get outta there and go home.

THE HIGH HUM LEVEL— FM BROADCASTING STATION

During that same period I worked part-time for an AM/FM broadcaster. They were mostly country music in the past, but had just started carrying what passed for "folk music" in those days. FM broadcasting was relatively new, and only then were large numbers of FM broadcast receivers being installed on hi-fi sets. In previous times, FM receivers were add-ons to AM designs, so tended to be low-fi. No one noticed the 60-Hz hum that permeated the signal because their receivers' audio rolled off considerably above 60 Hz (the −3 dB point was usually about 200 to 300 Hz). But when Dick Cerri's Music Americana went on the air, a lot of listeners called in and complained. The audience for that show had a lot of hi-fi owners . . . whose

equipment worked really well at 60 Hz. That hum was a huge component of the signal.

The problem with that system was that someone managed to install a new studio and never even considered using a common ground. One night, the chief engineer showed up with a roll of copper roofing flashing (7 inches wide, 1-lb/ft^2), a foot-square 1/4-inch copper plate, some tinned copper braid . . . and a drill. We placed the copper plate underneath the disk jockey's desk and ran bonding braid from all pieces of equipment to that plate. The copper flashing was routed down the back of the desk, under the wall (it was fake wall), to the transmitter. The flashing was bolted to a grounding surface on the footers of the transmitter. Unfortunately, there were no connectors and I got to break three bits on some of the hardest steel I've seen trying to fasten that darn copper flashing.

When the chief engineer measured the hum before and after, we were certain to have achieved at least 50 dB of suppression. It might've been more, but that was the limit of our test equipment.

THE SLACK COAX CAPER

Some years later I was at Old Dominion College in Norfolk, Virginia, and working part-time to pay my way. Another rough and ready fellow was named Dexter. He was a fellow ham radio operator (which is how I met him), but he was also a broadcast engineer for one of the larger independent AM broadcast stations in the area. In his part time ol' Dexter would found a new FM broadcast station, get the license, rent it and his station in his garage to a budding new broadcaster, and sit back and collect the money. The broadcaster would stay in Dexter's garage until they could build their own transmitter and get it FCC approved. I was rather amused when someone showed me a copy of a book by televangelist Pat Robertson that showed the Christian Broadcasting Network's first FM station . . . it was a clear picture of Dexter's garage.

One day, I was riding with another ham operator on the way over to Dexter's house.

We were listening to another station, but were soon pretty certain that the interference we were hearing was coming from Dexter's transmitter. It was all up and down the FM band! Every half megahertz or so, there was Dexter's FM signal. The signal itself was really broad on its own frequency. When we got over to the house we ran inside and told ol' Dexter what we heard. "[Darn!]"—not his real word—"the shoestring slipped again." Huh? The *what* slipped?

We followed Dexter out to the garage and watched him open the rear door of the transmitter. Oddly, the thing didn't go off the air (the door AC interlocks were all shorted! Don't try that at home, kiddies!). The deviation meter was slashing back and forth, rather than oscillating about a fixed (legal) point. The 1,000-watt final amplifier was in one deck of the 19-inch rack, and the exciter and modulator were in another deck. The coaxial cable between them had a black piece of shoelace dangling free. Dexter pushed the coax over to the side of the cab-

inet and refastened the knot that secured it. Sure enough, the deviation meter settled down, and a quick check on my buddy's car radio showed the problem cured . . . even though the car radio was close enough to the antenna to be overloaded. For all these years I haven't been able to figure out why tying off the coax that way stopped the oscillation.

LESSON LEARNED: DON'T COMPLAIN TOO LOUDLY

My next example was one that I handled for a customer of a radio shop, but also as a ham radio operator. A local ham operator was accused of causing television interference. His neighbors could hear his voice on their sets. In our area, the local FCC Field Engineering Office relied on volunteers to solve TVI problems for commercial, amateur radio, and CB stations, and I was one of the volunteers. Consider Figure 15.1. The interfering signal

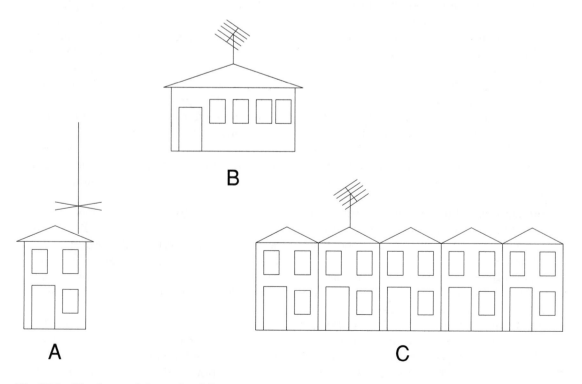

Fig. 15.1 *"Don't complain too loudly" scenario.*

was located in a duplex house ("A" in Figure 15.1), and it had a vertical antenna.

Although the guy was a ham operator (whom I knew personally), he was actually using a citizens band set. Because he was operating the CB legally, he had only 4 watts or so of RF output. It was the CB set that was causing the problem. Four watts? You gotta be kidding. I checked the output of the transmitter, and it was right at 4 watts. I then inserted a pretty heavy duty low-pass filter with a 33-MHz cutoff frequency and a very sharp roll-off slope. The output of the CB set went down only a very small amount equal to the insertion loss of the filter. That's a good sign that there was no significant harmonic radiation (a spectrum analyzer would be better!). Other tests showed that there were no spurious emissions of any sort. Yet the interference persisted even with the filter in-line. Grounding was ruled out.

Figure 15.1 shows the approximate geometry. The buildings were separated by not more than 50 yards or so. The CB rig was at "A," and the complainants were at "B" (a single family home), and "C" (townhouse block). One of the loudest and most profane of the complainers was the president of the townhouse homeowners' association. The guy was a real jerk.

My first attempt was to solder a high-pass TVI filter directly to the antenna terminals of the tuner inside the TV set at "B" (I was also a qualified consumer electronics Certified Electronics Technician). The problem persisted. The interference didn't even abate. That's not supposed to happen, by the way. Putting a low-pass filter on the HF transmitter and a high-pass filter directly on the TV tuner front end is supposed to nip the problem in the bud. Right? After all, the ARRL Handbook said so.

Not if the interfering signal is on-channel! So how could that be? The emission from the transmitter at "A" was at 27 MHz or so, and the interference was to VHF television channels, some of which weren't harmonically related to the transmitter frequency. I borrowed a Stoddard Field Strength Meter and went to work looking for the source of the

problem. Sure enough, a signal was present on one of the non-harmonically related TV channels . . . and it got stronger and stronger as I got closer to the townhouse block ("C").

The problem turned out to be quite simple. The townhouse block had one television antenna in the center of the cluster (antennas were frowned on, and cable hadn't been installed in that neighborhood), so they used a Master Antenna TV (MATV) system. A single high-gain antenna served the entire block. To give it enough signal, there was a 60-dB wideband amplifier at the antenna "head end." My more physically fit buddy climbed into the attic of the townhouse cluster and turned off the amplifier, while the CBer at "A" was transmitting. Simultaneously, the interference at "B" also disappeared!

We later pieced together what happened. The front-end RF amplifier transistor (a PNP germanium unit) was leaky (weren't most of 'em in those days!), and it was easily saturated. When the CB signal was picked up on the twin-lead transmission line (they didn't even use coax from the antenna to the amplifier!), it was rectified by the RF amplifier transistor. This created a large number of harmonics. To make matters more interesting, there were also a large number of TV and FM signals applied to the amplifier as well. These mixed together to produce a mish-mash of intermodulation products ($F = nF_1 \pm mF_2$) that were reradiated back out the TV antenna. Unfortunately, for many of the frequencies the antenna was not only resonant (making the reradiation very effective), it produced gain. The overall result was interference to both "B" and "C" sites.

I proved that the amplifier was causing the problem (and have since learned that was not a rare case!), but the president of the homeowners' association still complained that it must've been the CBer's 4 watts that wrecked the amplifier. You can't win with some people.

One lesson learned: Keep your blaming, rebuking mouth zipped until you know for sure where the fault lies . . . it might be in your own house (do I hear someone talking about people in glass townhouses?).

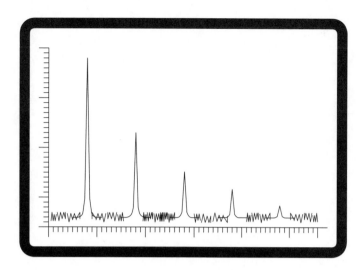

Fig. 15.2
Spectrum analyzer display.

When doing EMI troubleshooting on radio transmitters, look for spurious emissions (parasitics) and harmonics from a transmitter with a communications receiver, a field strength meter, or a tunable wavemeter (see Chapter 8). Today, we would probably use a spectrum analyzer, rather than a field strength meter. These instruments are essentially swept receivers with the output displayed on a cathode ray oscilloscope (CRO) as an amplitude-vs-frequency plot (Figure 15.2).

If the spectrum analyzer is used, then it becomes really easy to check the output of a transmitter to see if the harmonics are legal. If the rules call for a harmonic to be −40 dBc (decibels below the carrier), then it becomes immediately apparent on the spectrum analyzer if the spec is not being met. You can also see if any other signals are present, and do a site analysis to determine the possible combinations of signals. Once you work the $F = nF_1 \pm mF_2$ equation, you can tell something about the nature of interference (and what possibilities exist). You can also determine which frequencies to filter out far more effectively than the old guessing methods.

Spectrum analyzers can be quite expensive. It's possible to pay $40,000 for one. But today you can also buy commercially built spectrum analyzers for less than $2,000. Some of these work by using your oscilloscope as the display, while others have a built-in oscilloscope.

RUSTY DOWNSPOUTS

When rain gutters and downspouts were made of either copper or steel, rather than aluminum or plastic, there was a possibility that EMI could occur when the joint between the downspout and gutter corroded. The oxide layer formed a natural PN junction with the metal, so would rectify any RF signals that appeared on the downspout or gutter. Those pieces were long enough to pick up signals at relatively low frequencies. Some people claim this effect was seen quite frequently, but it may be in the realm of technomyth. Keep it in mind, however, whenever you see harmonics that are difficult to track down. It may be true. Corroded items in the vicinity of any transmitting antenna could be a culprit.

Chapter 16

Radio Receiver Basics

Before one can deal with EMI to radio receiver systems, it helps to understand a little about how such systems work. From that information, strategies for EMI mitigation and elimination can be devised. In this chapter you will find the basic theories of radio receiver design, their specifications, and how EMI can affect the system.

SIGNALS, NOISE, AND RECEPTION

No matter how simple or fancy the system may be, the basic function of a radio receiver is the same: to distinguish signals from noise. The concept "noise" covers both man-made and natural radio-frequency signals. The man-made signals include all signals in the passband other than the one being sought.

In communications systems the signal is some form of modulated (AM, FM, PM, etc.) periodic sine wave propagating as an electromagnetic (i.e., radio) wave. The "noise," on the other hand, tends to be a random signal that sounds like the "hiss" heard between stations on a radio. The spectrum of such noise signals appears to be Gaussian ("white noise") or pseudo-Gaussian ("pink noise" or "bandwidth limited" noise).

In radio astronomy and satellite communications systems the issue is complicated because the signals are also noise. The radio emissions of Jupiter and the Sun are very much like the Gaussian or pseudo-Gaussian signals that are, in other contexts, nothing but useless noise. In fact, in the early days of radar the galactic noise tended to mask returns from incoming enemy aircraft, so to the radar operators these signals were noise of the worst kind. Yet to a radio astronomer, those signals are the goal! In satellite communications systems the "signals" of the radio astronomer are limitations and annoyances at best and devastating at worst. The trick is to separate out the noise you want from the noise you don't.

Figure 16.1A shows an amplitude-vs-time plot of a typical noise signal, while Figure 16.1B shows a type of regular radio signal that could be generated by a transmitter. Notice the difference between the two. The signal is regular and predictable. Once you know the frequency and period you can predict the amplitude at other points along the

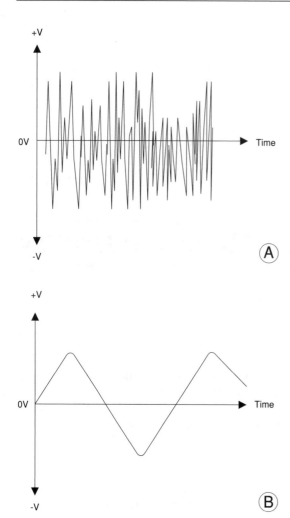

Fig. 16.1   *(A) Noise signal; (B) signal.*

time line. The noise signal, on the other hand, is unpredictable. Knowing the cycle-to-cycle amplitude and duration (there is no true "period") does not confer the ability to predict anything at all about the following cycles.

   In some receivers, especially those designed for pulse reception, the differences highlighted between Figures 16.1A and 16.1B are used to increase the performance of the receiver. An integrator circuit finds the time average of the input signal. True Gaussian noise integrated over a sufficiently long term will average to zero. This occurs because Gaussian noise contains all phases, ampli-

tudes, and polarities randomly distributed. Pseudo-Gaussian noise is bandwidth limited, so it may not integrate to zero, but very near it. The signal, on the other hand, will integrate to some nonzero value, and so will stand out in the presence of integrated noise.

THERMAL NOISE

Every electronic system (even a simple resistor) generates thermal noise, even if there is no power flowing in it. One of the goals of the system designer is to minimize the noise added by the system, so that weaker signals are not obscured. One of the basic forms of noise seen in systems is the thermal noise. Even if the amplifiers in the receiver add no noise (they will!), there will be thermal noise at the input due to the input resistance.

   If you replace the antenna with a resistor matched to the system impedance that is totally shielded, there will still be noise present. The noise is produced by the random motion of electrons inside the resistor. At all temperatures above absolute zero (about −273.16°C) the electrons in the resistor material are in random motion. At any given instant there will be a huge number of electrons in motion in all directions. The reason why there is no discernible current flow in one direction is that the motions cancel each other out even over short time periods. The noise power present in a resistor is:

$$P_N = KTBR \text{ watts} \qquad (16.1)$$

where:

   P_N is the noise power in watts
   T is the temperature in kelvins (K)
   K is Boltzmann's constant (1.38×10^{-23}
      joules/K)
   R is the resistance in ohms (Ω)

NOTE:
By international agreement T is set to 290 K

   Consider a receiver with a 1-MHz bandwidth and an input resistance of 50 ohms. The noise power is $(1.38 \times 10^{-23}$ joules/K$) \times (290$ K$)$ $\times (1,000,000$ Hz$) \times (50$ $\Omega) = 2 \times 10^{-13}$ watts.

THE RECEPTION PROBLEM

Figure 16.2 shows the basic problem of radio reception, especially in cases where the signal is very weak. The signal in Figure 16.2A is embedded in noise that is relatively high amplitude. This signal is lower than the noise level and so is very difficult (perhaps impos-

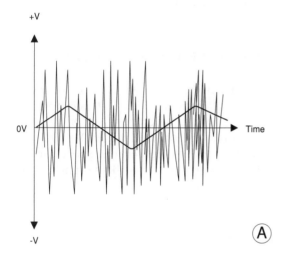

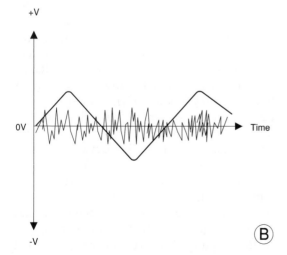

Fig. 16.2 *(A) Signal and noise combined;*
(B) reduced noise makes signal accessible.

sible) to detect. The signal in Figure 16.2B is easily detectable because the signal amplitude is higher than the noise amplitude. It becomes difficult when the signal is only slightly stronger than the average noise power level.

The signal-to-noise ratio (SNR) of a receiver system tells us something about the detectability of the signal. The SNR is normally quoted in *decibels* (dB), which are defined as:

$$\text{SNR} = 10 \log \left[\frac{P_S}{P_N} \right] \text{dB} \qquad (16.2)$$

where:

SNR is the signal-to-noise ratio in decibels (dB)
P_S is the signal power level
P_N is the noise power level

How high an SNR is required? That depends on a lot of subjective factors when a human listener is present. Skilled radio operators can detect signals with an SNR of less than 1 dB . . . but the rest of us cannot even hear that signal. Most radio operators can detect 3-dB SNR signals, but for "comfortable" listening 10-dB SNR is usually specified. For digital systems the noise performance is usually defined by the acceptable bit error rate (BER).

STRATEGIES

A number of strategies can be used to improve the SNR of a system. First, of course, is to buy a receiver that has a low internal "noise floor" and not do anything to upset that figure. High-quality receivers have very low noise, but there is sometimes some creative spec writing in the advertisements—where different bandwidths are used for the measurement, and only the most favorable value (which may not be the bandwidth that matches your needs) is reported.

By common sense we see that there are two approaches to SNR improvement: either increase the signal amplitude or decrease the noise amplitude. Most successful systems do both, but it must be done carefully.

One approach to SNR improvement is to use a preamplifier ahead of the receiver antenna terminals. This approach may or may not work, and under some situations may make the situation worse. The problem is that the preamplifier adds noise of its own and will amplify noise from outside (received through the antenna) and the desired signal equally. If you have an amplifier with a gain of, say, 20 dB, then the external noise is increased by 20 dB and the signal is increased by 20 dB. The result is that the absolute numbers are bigger but the SNR is the same. If the amplifier produces any significant noise of its own, then the SNR will degrade. The key is to use a very low noise amplifier (LNA) for the preamplifier. Using an LNA for the preamplifier may actually reduce the noise figure of the receiver system.

Another trick is to use a preselector ahead of the receiver. A preselector is either a tuned circuit or a bandpass filter placed in the antenna transmission line ahead of the receiver antenna terminals. A passive preselector has no amplification (uses L-C elements only), while an active preselector has a built-in amplifier. The amplifier should be an LNA type. The reason why the preselector can improve the system is that it amplifies the signal by a fixed amount, but only the noise within the passband is amplified the same amount as the signal. Improvement comes from bandwidth limiting the noise but not the signal.

Another practical approach is to use a directional antenna. This method works especially well when the unwanted noise is other man-made signal sources. An omnidirectional antenna receives equally well in all directions. As a result, both natural and man-made external noise sources operating within the receiver's passband will be picked up. But if the antenna is made highly directional, then all noise sources that are not in the direction of interest are suppressed.

Highly directional antennas have gain, so the signal levels in the direction of interest are increased. Although the noise also increases in that direction, the rest of the noise sources (in other directions) are suppressed. The result is that SNR is increased by both methods.

When designing a communications system, the greatest attention should usually be paid to the antenna, then to an LNA or low-noise preselector, and then to the receiver. Generally speaking, money spent on the antenna gives more SNR for a given investment than the same money spent on amplifiers and other attachments.

RADIO RECEIVER SPECIFICATIONS

Radio receivers are at the heart of nearly all communications activities. In this chapter we will discuss the different types of radio receivers that are on the market. We will also learn how to interpret receiver specifications. Later, we will look at specific designs for specific applications.

ORIGINS

The very earliest radio receivers were not receivers at all, in the sense that we know the term today. Early experiments by Hertz, Marconi, and others used spark gaps and regular telegraph instruments of the day. Range was severely limited because those devices have a terribly low sensitivity to radio waves. Later, around the turn of the 20th century, a device called a Branly coherer was used for radio signal detection. This device consisted of a glass tube filled with iron filings placed in series between the antenna and ground. Although considerably better than earlier apparatus, the coherer was something of a dud for weak signal reception. In the first decade of this century, however, Fleming invented the diode vacuum tube, and Lee DeForest invented the triode vacuum tube. These devices made amplification possible and detection a lot more efficient.

A receiver must perform two basic functions: (1) It must respond to, detect, and demodulate desired signals; and (2) It must not respond to, detect, or be adversely affected by undesired signals. If it fails in either of these two functions, then it is a poorly performing design.

Both functions are necessary. Weakness in either function makes a receiver a poor bargain unless there is some mitigating circumstance. The receiver's performance specifications tell us how well the manufacturer claims that their product does these two functions.

Crystal Video Receivers

Crystal video receivers (Figure 16.3) grew out of primordial crystal sets, but are sometimes used in microwave bands even today. The original crystal sets (pre-1920) used a naturally occurring PN junction "diode" made from a natural lead compound called galena crystal with an inductor–capacitor (L-C) tuned circuit. Later, crystal sets were made using germanium or silicon diodes. When vacuum tubes became generally available, it

was common to place an audio amplifier at the output of the crystal set. Modern crystal video receivers use silicon or gallium-arsenide microwave diodes and a wideband video amplifier (rather than the audio amplifier). Applications include some speed radar receivers, aircraft warning receivers, and some communications receivers (especially short-range).

Tuned Radio Frequency (TRF) Receivers

The tuned radio frequency (TRF) radio receiver uses an L-C resonant circuit in the front end, followed by one or more radio frequency amplifiers ahead of a detector stage. Two varieties are shown in Figures 16.4 and 16.5. The version in Figure 16.4 is called a tuned gain-block receiver. It is commonly used in monitoring very low frequency (VLF) signals to detect solar flares and sudden ionospheric disturbances (SIDs). Later versions of the TRF concept use multiple tuned RF circuits between the amplifier stages. These designs are also used in VLF solar flare/SID monitoring. Early models used independently tuned L-C circuits, but those proved to be very difficult to tune without creating an impromptu Miller oscillator circuit. Later versions mechanically

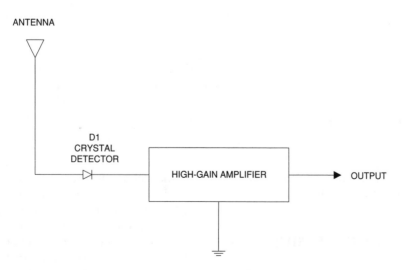

Fig. 16.3 Crystal video receiver.

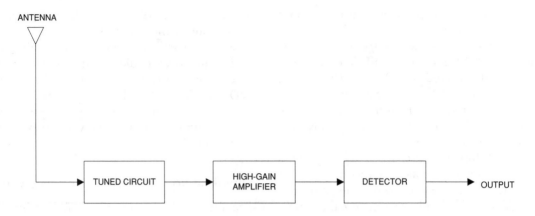

Fig. 16.4 *Simple tuned radio frequency (TRF) receiver.*

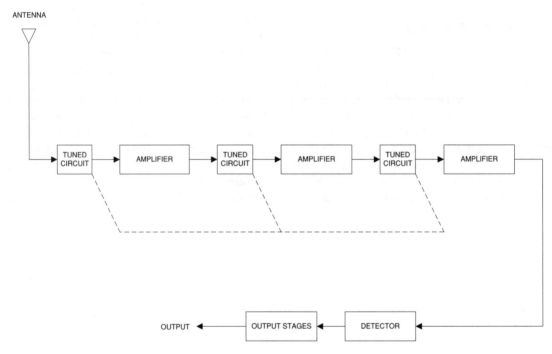

Fig. 16.5 *Complex TRF radio receiver.*

linked ("ganged") the tuned circuits to operate from a single tuning knob.

SUPERHETERODYNE RECEIVERS

Figure 16.6 shows the block diagram of a superheterodyne receiver. We will use this hypothetical receiver as the basic generic framework for evaluating receiver performance. The design in Figure 16.6 is called a superheterodyne receiver and represents the largest class of radio receivers; it covers the vast majority of receivers on the market.

The superheterodyne receiver block diagram of Figure 16.6 is typical of many re-

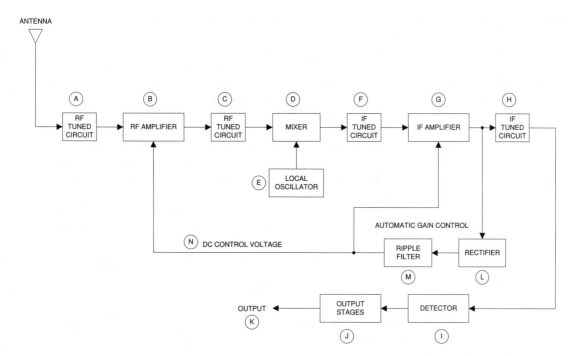

Fig. 16.6 *Superheterodyne receiver.*

ceivers. The purpose of a superheterodyne is to convert the incoming RF frequency to a single frequency where most of the signal processing takes place. The front-end section of the receiver consists of the *radio-frequency* (RF) amplifier and any RF tuning circuits that may be used (A-B-C in Figure 16.6). In some cases, the RF tuning is very narrow and basically tunes one frequency. In other cases, the RF front-end tuning is broadbanded. In that case, bandpass filters are used.

The frequency translator section (D and E) is also considered part of the front end in most textbooks, but here we will label it as a separate entity. The translator consists of a frequency mixer and a local oscillator. This section does the heterodyning, which is discussed in more detail below. The output of the frequency translator is called the intermediate frequency (IF).

The translator stage is followed by the intermediate frequency amplifier. The IF amplifier (F-G-H) is basically a radio-frequency amplifier tuned to a single frequency. The IF can be higher or lower than the RF frequency, but it will always be a single frequency.

A sample of the IF amplifier output signal is applied to an automatic gain control (AGC) section (L-M). The purpose of this section is to keep the signal level in the output more or less constant. The AGC circuit consists of a rectifier and ripple filter that produces a DC control voltage. The DC control voltage is proportional to the input RF signal level (N). It is applied to the IF and RF amplifiers to raise or lower the gain according to signal level. If the signal is weak, then the gain is forced higher, and if the signal is strong the gain is lowered. The end result is to smooth out variations of the output signal level.

The detector stage (I) is used to recover any modulation that is on the input RF signal. The type of detector depends on the type of modulation used for the incoming signal. Amplitude modulation (AM) signals are generally handled in an envelope detector. In some cases a special variant of the envelope detector called a square law detector is used. The difference is that the straight envelope detector is linear, while the square law detector is nonlinear. Single-sideband (SSB), double-sideband suppressed carrier (DSBSC), and keyed CW signals will use a product detector, while FM and PM need a frequency or phase sensitive detector.

The output stages (J-K) are used to amplify and deliver the recovered modulation to the user. If the receiver is for broadcast use, then the output stages are audio amplifiers and loudspeakers. In some radio astronomy and instrumentation telemetry receivers, the output stages consist of integrator circuits and DC amplifiers.

Heterodyning

The main attribute of the superheterodyne receiver is that it converts the radio signal's RF frequency to a standard frequency for further processing. Although today the new frequency, called the intermediate frequency or IF, may be either higher or lower than the RF frequencies, early superheterodyne receivers always down-converted RF signal to a lower IF frequency (IF < RF). The reason was purely practical, for in those days higher frequencies were more difficult to process than lower frequencies. Even today, because variable tuned circuits still tend to offer different performance over the band being tuned, converting to a single IF frequency and obtaining most of the gain and selectivity functions at the IF allows more uniform overall performance over the entire range being tuned.

A superheterodyne receiver works by frequency converting ("heterodyning"—the "super" part is 1920s vintage advertising hype) the RF signal. This occurs by nonlinearly mixing the incoming RF signal with a local oscillator (LO) signal. When this process is done, disregarding noise, the output spectrum will contain a large variety of signals according to:

$$F_O = mF_{RF} \pm nF_{LO} \qquad (16.3)$$

where:

F_{RF} is the frequency of the RF signal
F_{LO} is the frequency of the local oscillator
m and n are either zero or integers (0, 1, 2, 3 . . . . n)

Equation 16.3 means that there will be a large number of signals at the output of the mixer, although for the most part the only ones that are of immediate concern to understanding superheterodyne operation are those for which m and n are either 0 or 1. Thus, for

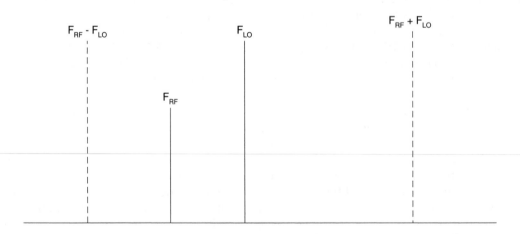

FREQUENCY

Fig. 16.7   Relationship of F_{RF}, F_{LO}, and the sum and difference frequencies.

our present purpose, the output of the mixer will be the fundamentals (F_{RF} and F_{LO}), and the second-order products ($F_{LO} - F_{RF}$ and $F_{LO} + F_{RF}$), as seen in Figure 16.7. Some mixers, notably those described as *double-balanced mixers* (DBMs), suppress F_{RF} and F_{LO} in the mixer output, so only the second-order sum and difference frequencies exist with any appreciable amplitude. This case is simplistic, and is used only for this present discussion. Later on, we will look at what happens when third-order ($2F_1 \pm F_2$ and $2 F_2 \pm F_1$) and fifth-order ($3 F_1 \pm 2 F_2$ and $3 F_2 \pm 2 F_1$) frequencies become large.

Note that the local oscillator frequency can be either higher than the RF frequency (high-side injection) or lower than the RF frequency (low-side injection). There is ordinarily no practical reason to prefer one over the other except that it will make a difference whether an analog main tuning dial (if used) reads high-to-low or low-to-high.

The candidates for IF are the sum (LO + RF) and difference (LO – RF) second-order products found at the output of the mixer. A high-Q tuned circuit following the mixer will select which of the two are used. Consider an example. Suppose an AM broadcast band superhet radio has an IF frequency of 455 kHz and the tuning range is 540 to 1,700 kHz. Because the IF is lower than any frequency within the tuning range, it will be the difference frequency that is selected for the IF. The local oscillator is set to be high-side injection, so will tune from (540 + 455) = 995 kHz, to (1,700 + 455) = 2,155 kHz.

Front-End Circuits

The principal task of the front-end and frequency translator sections of the receiver in Figure 16.6 is to select the signal and convert it to the IF frequency. But in many radio receivers there may be additional functions. In some cases (but not all), an RF amplifier will be used ahead of the mixer. Typically, these amplifiers have a gain of 3 to 10 dB, with 5 to 6 dB being very common. The tuning for the RF amplifier is sometimes a broad bandpass fixed-frequency filter that admits an entire band. In other cases, it is a narrowband, but variable-frequency, tuned circuit.

Intermediate Frequency (IF) Amplifier

The IF amplifier is responsible for providing most of the gain in the receiver, as well as the narrowest bandpass filtering. It is a high-gain, usually multistaged, single-frequency tuned radio frequency amplifier. For example, one HF shortwave receiver block diagram lists 120 dB of gain from antenna terminals to audio output, of which 85 dB are provided in the 8.83-MHz IF amplifier chain. In the example of Figure 16.6, the receiver is a single conversion design, so there is only one IF amplifier section.

Detector

The detector demodulates the RF signal and recovers whatever audio (or other information) is to be heard by the listener. In a straight AM receiver, the detector will be an ordinary half-wave rectifier and ripple filter and is called an envelope detector. In other detectors, notably double-sideband suppressed carrier (DSBSC), single-sideband suppressed carrier (SSBSC or SSB), or continuous wave (CW or Morse telegraphy), a second local oscillator, usually called a beat frequency oscillator (BFO), operating near the IF frequency, is heterodyned with the IF signal. The resultant difference signal is the recovered audio. That type of detector is called a product detector. Many AM receivers today have a sophisticated synchronous detector, rather than the simple envelope detector.

Audio Amplifiers

The audio amplifiers are used to finish the signal processing. They also boost the output of the detector to a usable level to drive a loudspeaker or set of earphones. The audio amplifiers are sometimes used to provide additional filtering. It is quite common to find narrowband filters to restrict audio bandwidth, or notch filters to eliminate interfering signals that make it through the IF amplifiers intact.

RECEIVER PERFORMANCE FACTORS

There are three basic areas of receiver performance that must be considered. Although interrelated, they are sufficiently different to merit individual consideration: noise, static attributes, and dynamic attributes. We will look at all of these areas, but first let's look at the units of measure that we will use in this series.

UNITS OF MEASURE

Input Signal Voltage

Input signal level, when specified as a voltage, is typically stated in either microvolts (μV) or nanovolts (nV). The volt is simply too large a unit for practical use on radio receivers. Signal input voltage (or sometimes power level) is often used as part of the sensitivity specification, or as a test condition for measuring certain other performance parameters.

There are two forms of signal voltage that are used for input voltage specification: source voltage (V_{EMF}) and potential difference (V_{PD}), as illustrated in Figure 16.8. The source voltage (V_{EMF}) is the open-terminal (no-load) voltage of the signal generator or source, while the potential difference (V_{PD}) is the volt-

age that appears across the receiver antenna terminals with the load connected (the load is the receiver antenna input impedance, R_{in}). When $R_s = R_{in}$, the preferred "matched impedances" case in radio receiver systems, the value of V_{PD} is one-half V_{EMF}. This can be seen in Figure 16.8 by noting that R_s and R_{in} form a voltage divider network driven by V_{EMF}, and with V_{PD} as the output.

dBm

These units refer to decibels relative to one milliwatt (1 mW) dissipated in a 50-ohm resistive impedance (defined as the 0-dBm reference level), and is calculated from:

$$dBm = 10 \log \left[\frac{P_{\text{Watts}}}{0.001} \right] \qquad (16.4)$$

or

$$dBm = 10 \log (P_{MW}). \qquad (16.5)$$

In the noise voltage case calculated above, 0.028 μV in 50 ohms, the power is $V^2/50$, or 5.6×10^{-10} watts, which is 5.6×10^{-7} mW. In dBm notation, this value is 10 log (5.6×10^{-7}), or –62.5 dBm.

dBmV

This unit is used in television receiver systems in which the system impedance is 75 ohms, rather than the 50 ohms normally used in other RF systems. It refers to the signal voltage, measured in decibels, with respect to a signal level of one millivolt (1 mV) across a 75-ohm resistance (0 dBmv). In many TV specs, 1 mV

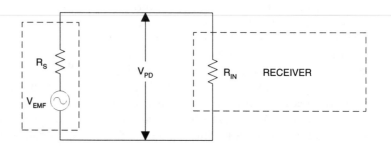

Fig. 16.8
Relationship of V_{EMF} and V_{PD}.

is the full quieting signal that produces no "snow" (i.e., noise) in the displayed picture. Note: 1 mV = 1,000 µV.

dBµV

This unit refers to a signal voltage, measured in decibels, relative to one microvolt (1 µV) developed across a 50-ohm resistive impedance (0 dBµV). For the case of our noise signal voltage, the level is 0.028 µV, which is the same as −31.1 dBµV. The voltage used for this measurement is usually the V_{EMF}, so to find V_{PD} divide it by two after converting dBµV to µV.

RULE OF THUMB: To convert dBµV to dBm, subtract 113 dB; i.e., 100 dBµV = (100 dBµV − 113 dB) = −13 dBm.

It requires only a little algebra to convert signal levels from one unit of measure to another. This job is sometimes necessary when a receiver manufacturer mixes methods in the same specifications sheet. In the case of dBm and dBµV, 0 dBµV is 1 µV V_{EMF}, or a V_{PD} of 0.5 µV, applied across 50 ohms, so the power dissipated is 5×10^{-15} watts, or −113 dBm.

NOISE

A radio receiver must detect signals in the presence of noise. The signal-to-noise ratio is the key here because a signal must be above the noise level before it can be successfully detected and used.

Noise comes in a number of different guises, but for sake of this discussion we can divide them into two classes: sources external to the receiver and sources internal to the receiver. There is little one can do about the external noise sources, for they consist of natural and man-made electromagnetic signals that fall within the passband of the receiver. Figure 16.9A shows an approximation of the external noise situation from the middle of the AM broadcast band to the low end

of the VHF region. A somewhat different view, which captures the severe noise situation seen by receivers, is shown in Figure 16.9B. One must select a receiver that can cope with external noise sources, especially if the noise sources are strong.

Some natural external noise sources are extraterrestrial. It is these signals that form the basis of radio astronomy. For example, if you aim a beam antenna at the eastern horizon prior to sunrise, a distinct rise of noise level occurs as the Sun slips above the horizon, especially in the VHF region (the 150 to 152 MHz band is used to measure solar flux). The reverse occurs in the west at sunset, but is less dramatic, probably because atmospheric ionization decays much more slowly than it is generated. During World War II, it is reported that British radar operators noted an increase in received noise level any time the Milky Way was above the horizon, decreasing the range at which they could detect in-bound German bombers. There is also some well-known, easily observed noise from the planet Jupiter in the 18- to 30-MHz frequency range.

The receiver's internal noise sources are determined by the design of the receiver. Ideal receivers produce no noise of their own, so the output signal from the ideal receiver would contain only the noise that was present at the input along with the radio signal. But real receiver circuits produce a certain level of internal noise of their own. Even a simple fixed-value resistor is noisy. Figure 16.10A shows the equivalent circuit for an ideal, noise-free resistor, while Figure 16.10B shows a practical real-world resistor. The noise in the real-world resistor is represented in Figure 16.10B by a noise voltage source, V_n, in series with the ideal, noise-free resistance, R_i. At any temperature above absolute zero (0 K or about −273°C) electrons in any material are in constant random motion. Because of the inherent randomness of that motion, however, there is no detectable current in any one direction. In other words, electron drift in any single direction is cancelled over even short time periods by equal

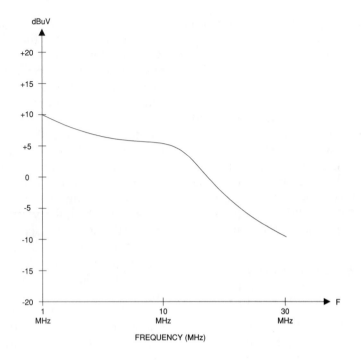

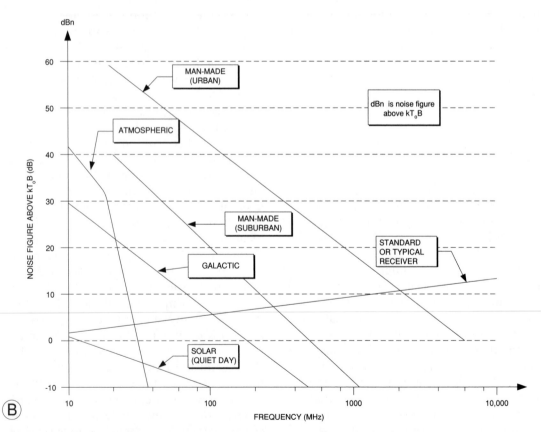

Fig. 16.9 *(A) External noise situation of receiver; (B) all noise sources.*

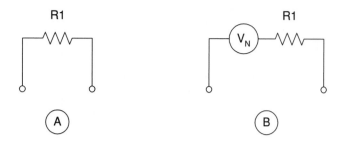

Fig. 16.10
(A) Ideal resistor; (B) real resistor with noise source.

drift in the opposite direction. Electron motions are therefore statistically decorrelated. There is, however, a continuous series of random current pulses generated in the material, and those pulses are seen by the outside world as noise signals.

If a perfectly shielded 50-ohm resistor is connected across the antenna input terminals of a radio receiver, the noise level at the receiver output will increase by a predictable amount over the short-circuit noise level. Noise signals of this type are called by several names: thermal agitation noise, thermal noise, or Johnson noise. This type of noise is also called "white noise" because it has a very broadband (nearly Gaussian) spectral density. The thermal noise spectrum is dominated by mid-frequencies (10^4 to 10^5 Hz) and is essentially flat. The term "white noise" is a metaphor developed from white light, which is composed of all visible color frequencies. The expression for such noise is:

$$V_n = \sqrt{4\,KTBR} \qquad (16.6)$$

where:

V_n is the noise potential in volts (V)
K is Boltzmann's constant (1.38×10^{-23} J/K)
T is the temperature in kelvins (K), normally set to 290 or 300 K by convention.
R is the resistance in ohms (Ω)
B is the bandwidth in hertz (Hz)

Table 16.1 and Figure 16.11 show noise values for a 50-ohm resistor at various bandwidths out to 5 and 10 kHz, respectively. Because different bandwidths are used for different reception modes, it is common practice to delete the bandwidth factor in Equation 16.6 and write it in the form:

$$V_n = \sqrt{4KTR}\ \text{V}\,/\sqrt{\text{Hz}} \qquad (16.7)$$

With Equation 16.7 one can find the noise voltage for any particular bandwidth by taking its square root and multiplying it by the equation. This equation is essentially the solution of the previous equation normalized for a 1-Hz bandwidth.

Table 16.1 Bandwidth vs Noise Voltage

Bandwidth (Hz)	Noise Voltage
1,000	2.83E-08
1,500	3.46E-08
2,000	4.00E-08
2,500	4.47E-08
3,000	4.90E-08
3,500	5.29E-08
4,000	5.66E-08
4,500	6.00E-08
5,000	6.33E-08
5,500	6.63E-08
6,000	6.93E-08
6,500	7.21E-08
7,000	7.49E-08
7,500	7.75E-08
8,000	8.00E-08
8,500	8.25E-08
9,000	8.49E-08
9,500	8.72E-08
10,000	8.95E-08

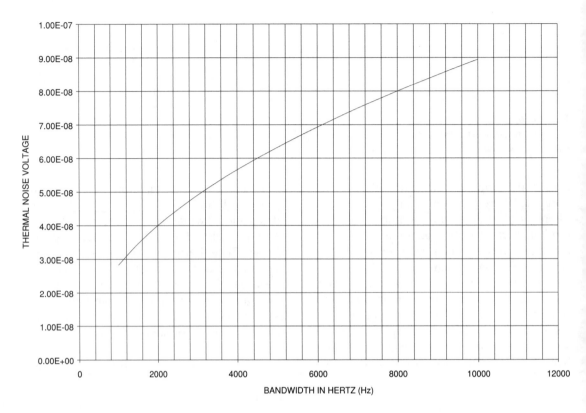

Fig. 16.11 *Thermal noise voltage vs bandwidth.*

SIGNAL-TO-NOISE RATIO (SNR OR S_n)

Receivers are evaluated for quality on the basis of signal-to-noise ratio (S/N or SNR), sometimes denoted S_n. The goal of the designer is to enhance the SNR as much as possible. Ultimately, the minimum signal level detectable at the output of an amplifier or radio receiver is that level which appears just above the noise floor level. Therefore, the lower the system noise floor, the smaller the minimum allowable signal.

NOISE FACTOR, NOISE FIGURE, AND NOISE TEMPERATURE

The noise performance of a receiver or amplifier can be defined in three different, but related, ways: noise factor (F_n), noise figure (NF), and equivalent noise temperature (T_e);

these properties are definable as a simple ratio, decibel ratio, or kelvin temperature, respectively.

Noise Factor (F_n)

For components such as resistors, the noise factor is the ratio of the noise produced by a real resistor to the simple thermal noise of an ideal resistor. The noise factor of a radio receiver (or any system) is the ratio of output noise power (P_{no}) to input noise power (P_{ni}):

$$F_n = \left[\frac{P_{no}}{P_{ni}} \right] T = 290\,\text{K} \qquad (16.8)$$

In order to make comparisons easier, the noise factor is usually measured at the standard temperature (T_o) of 290 K (standardized room temperature); although in some countries 299 K or 300 K are commonly used (the differences are negligible). It is also possible to define noise factor F_n in terms of the output and input signal-to-noise ratios:

$$F_n = \frac{S_{ni}}{S_{no}} \qquad (16.9)$$

where:

S_{ni} is the input signal-to-noise ratio
S_{no} is the output signal-to-noise ratio

Noise Figure (NF)

The noise figure is frequently used to measure the receiver's "goodness," i.e., its departure from "ideality." Thus, it is a figure of merit. The noise figure is the noise factor converted to decibel notation:

$$NF = 10 \log(F_n) \qquad (16.10)$$

where:

NF is the noise figure in decibels (dB)
F_n is the noise factor
log refers to the system of base-10 logarithms

Noise Temperature (T_e)

The noise "temperature" is a means for specifying noise in terms of an equivalent noise temperature—that is, the noise level that would be produced by a matching resistor (e.g., 50 ohms) at that temperature (expressed in kelvins). Evaluating the noise equations shows that the noise power is directly proportional to temperature in kelvins, and also that noise power collapses to zero at absolute zero (0 K).

Note that the equivalent noise temperature T_e is *not* the physical temperature of the amplifier, but rather a theoretical construct that is an equivalent temperature that produces that amount of noise power in a resistor. The noise temperature is related to the noise factor by:

$$T_e = (F_n - 1)T_o \qquad (16.11)$$

and to noise figure by:

$$T_e = KT_o \log^{-1}\left[\frac{NF}{10}\right] - 1. \qquad (16.12)$$

Noise temperature is often specified for receivers and amplifiers in combination with, or in lieu of, the noise figure.

NOISE IN CASCADE AMPLIFIERS

A noise signal is seen by any amplifier following the noise source as a valid input signal. Each stage in the cascade chain (Figure 16.12) amplifies both signals and noise from previous stages, and also contributes some additional noise of its own. Thus, in a cascade amplifier the final stage sees an input signal that consists of the original signal and noise amplified by each successive stage plus the noise contributed by earlier stages. The overall noise factor for a cascade amplifier can be calculated from Friis' noise equation:

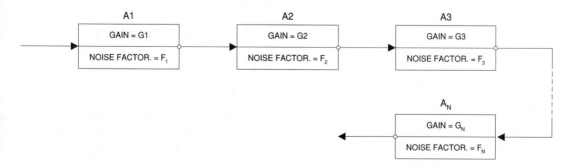

Fig. 16.12 **N-stage cascade system.**

$$F_n = F_1 + \frac{F_2 - 1}{G1} + \frac{F_3 - 1}{G1\,G2} + \dots$$

$$+ \frac{F_n - 1}{G1\,G2\dots G_{n-1}} \qquad (16.12)$$

where:

> F_n is the overall noise factor of N stages in cascade
> F_1 is the noise factor of stage-1
> F_2 is the noise factor of stage-2
> F_n is the noise factor of the nth stage
> $G1$ is the gain of stage-1
> $G2$ is the gain of stage-2
> G_{n-1} is the gain of stage $(n - 1)$

As you can see from Friis' equation, the noise factor of the entire cascade chain is dominated by the noise contribution of the first stage or two. High-gain, multistage RF amplifiers typically use a low-noise amplifier (LNA) circuit for the first stage or two in the cascade chain. Thus, you will find an LNA at the feedpoint of a satellite receiver's dish antenna, and possibly another one at the input of the receiver module itself, but other amplifiers in the chain might be more modest without harming system performance.

The matter of signal-to-noise ratio is sometimes treated in different ways that each attempt to crank some reality into the process. The signal-plus-noise-to-noise ratio ($S+N/N$) is found quite often. As the ratios get higher, the S/N and $S+N/N$ converge (only about 0.5 dB difference at ratios as little as 10 dB). Still another variant is the SINAD (signal-plus-noise-plus-distortion-to-noise) ratio. The SINAD measurement takes into account most of the factors that can deteriorate reception.

RECEIVER NOISE FLOOR

The noise floor of the receiver is a statement of the amount of noise produced by the receiver's internal circuitry and directly affects the sensitivity of the receiver. The noise floor is typically expressed in dBm. The noise

floor specification is evaluated as follows: the more negative the better. The best receivers have noise floor numbers of less than −130 dBm, while some very good receivers offer numbers of −115 dBm to −130 dBm.

The noise floor is directly dependent on the bandwidth used to make the measurement. Receiver advertisements usually specify the bandwidth, but be careful to note whether or not the bandwidth that produced the very good performance numbers is also the bandwidth that you'll need for the mode of transmission you want to receive. If, for example, you are interested only in weak 6-kHz wide AM signals, and the noise floor is specified for a 250-Hz CW filter, then the noise floor might be too high for your use.

STATIC MEASURES OF RECEIVER PERFORMANCE

The two principal static levels of performance for radio receivers are sensitivity and selectivity. The sensitivity refers to the level of input signal required to produce a usable output signal (variously defined). The selectivity refers to the ability of the receiver to reject adjacent channel signals (again, variously defined). Let's take a look at both of these factors. Keep in mind, however, that in modern high-performance radio receivers the static measures of performance, although frequently cited, may also be the least relevant compared with the dynamic measures (especially in environments with high interference levels).

SENSITIVITY

Sensitivity is a measure of the receiver's ability to pick up ("detect") signals, and is often specified in microvolts (μV). A typical specification might be "0.5-μV sensitivity." The question to ask is, "Relative to what?" The sensitivity number in microvolts is meaningless unless the test conditions are specified. For most commercial receivers, the usual test condition is the sensitivity required to produce a 10-dB signal-plus-noise-to-noise ($S+N/N$) ratio in the mode of interest. For example, if only one sensitivity figure is given, one must find out what bandwidth is being used: 5 to 6 kHz for AM, 2.6 to 3 kHz for single-sideband, 1.8 kHz for radioteletype, or 200 to 500 Hz for CW.

Indeed, one of the places where "creative spec writing" takes place for commercial receivers is that the advertisements will enthusiastically cite the sensitivity for a narrow bandwidth mode (e.g., CW), while the other specifications are cited for a more commonly used wider bandwidth mode (e.g., SSB). In one particularly egregious example, an advertisement claimed a sensitivity number that was applicable to the 270-Hz CW mode only, yet the 270-Hz CW filter was an expensive option that had to be specially ordered separately!

The amount of sensitivity improvement is seen by evaluating some simple numbers. Recall that a claim of "x-μV" sensitivity refers to some standard such as "x-μV to produce a 10-dB signal-to-noise ratio in y-Hz band-

width." Consider the case where the main mode for a high-frequency (HF) shortwave receiver is AM (for international broadcasting), the sensitivity is 1.9 μV for 10 dB SNR, and the bandwidth is 5 kHz. If the bandwidth were reduced to 2.8 kHz for SSB, then the sensitivity improves by the square root of the ratio, or SQRT(5/2.8). If the bandwidth is further reduced to 270 Hz (i.e., 0.27 kHz) for CW, then the sensitivity for 10 dB SNR is SQRT(5/0.27). The 1.9 μV AM sensitivity therefore translates to 1.42 μV for SSB and 0.44 μV for CW. If only the CW version is given, then the receiver might be made to look a whole lot better than it is, even though the typical user may never use the CW mode (note differences in Figure 16.13).

The sensitivity differences also explain why weak SSB signals can be heard under conditions when AM signals of similar strength have disappeared into the noise, or why the CW mode has as much as 20 dB advantage over SSB, ceteris paribus.

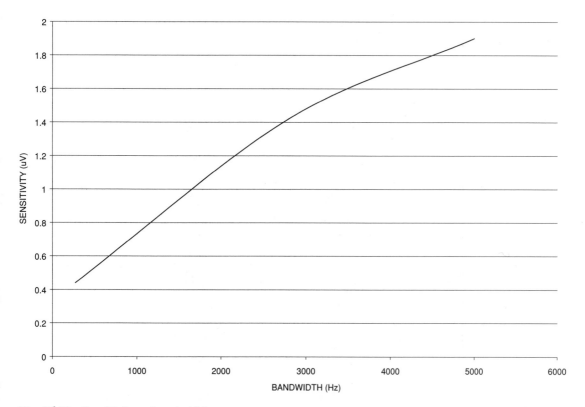

Fig. 16.13 *Sensitivity vs bandwidth.*

In some receivers, the difference in mode (AM, SSB, RTTY, CW, etc.) can conceivably result in sensitivity differences that are more than the differences in the bandwidths associated with the various modes. The reason is that there is sometimes a "processing gain" associated with the type of detector circuit used to demodulate the signal at the output of the IF amplifier. A simple AM envelope detector is lossy because it consists of a simple diode (1N60, etc.) and an R-C filter (a passive circuit without amplification). Other detectors (product detector for SSB, synchronous AM detectors) have their own signal gain, so they may produce better sensitivity numbers than the bandwidth suggests.

Another indication of sensitivity is minimum detectable signal (MDS), which is usually specified in dBm. This signal level is the signal power at the antenna input terminal of the receiver required to produce some standard $S+N/N$ ratio, such as 3 dB or 10 dB (Figure 16.14). In radar receivers, the MDS is usually described in terms of a single pulse return and a specified $S+N/N$ ratio. Also, in radar receivers, the sensitivity can be improved by integrating multiple pulses. If N return pulses are integrated, then the sensitivity is improved by a factor of N if coherent detection is used, and SQRT(N) if noncoherent detection is used.

Modulated signals represent a special case. For those sensitivities, it is common to specify the conditions under which the measurement is made. For example, in AM receivers the sensitivity to achieve 10 dB SNR is measured with the input signal modulated 30 percent by a 400 or 1,000 Hz sinusoidal tone.

An alternate method is sometimes used for AM sensitivity measurements, especially in servicing consumer radio receivers (where SNR may be a little hard to measure with the equipment normally available to technicians who work on those radios). This is the "standard output conditions" method. Some manuals will specify the audio signal power or

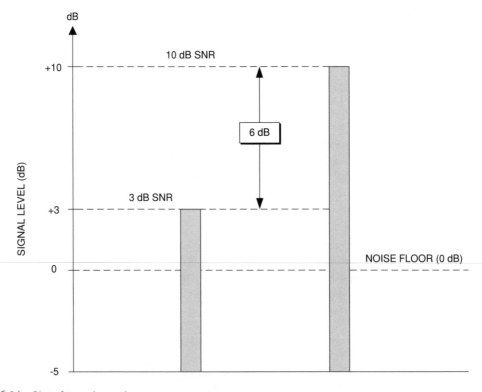

Fig. 16.14 Signal-to-noise ratio.

audio signal voltage at some critical point, when the 30 percent modulated RF carrier is present. In one automobile radio receiver, the sensitivity was specified as "X μV to produce 400 mW across 8 ohm resistive load substituted for the loudspeaker when the signal generator is modulated 30 percent with a 400 Hz audio tone." The cryptic note on the schematic showed an output sine wave across the loudspeaker with the label "400 mW in 8 Ω (1.79 volts), @30% mod. 400 Hz, 1 μV RF." What is missing is mention of the level of total harmonic distortion (THD) that is permitted.

The sensitivity is sometimes measured essentially the same way, but the signal levels will specify the voltage level that will appear at the top of the volume control, or output of the detector/filter, when the standard signal is applied. Thus, there are two ways seen for specifying AM sensitivity: 10 dB SNR and standard output conditions.

There are also two ways to specify FM receiver sensitivity. The first is the 10 dB SNR method discussed above, i.e., the number of microvolts of signal at the input terminals required to produce a 10-dB SNR when the carrier is modulated by a standard amount. The measure of FM modulation is deviation expressed in kilohertz. Sometimes, the full deviation for that class of receiver is used, while for others a value that is 25 to 35 percent of full deviation is specified.

The second way to measure FM sensitivity is the level of signal required to reduce the no-signal noise level by 20 dB. This is the 20-dB quieting sensitivity of the receiver. If you tune between signals on an FM receiver, you will hear a loud "hiss" signal, especially in the VHF/UHF bands. Some of that noise is externally generated, while some is internally generated. When an FM signal appears in the passband, that hiss is suppressed, even if the FM carrier is unmodulated. The quieting sensitivity of an FM receiver is a statement of the number of microvolts required to produce some standard quieting level, usually 20 dB.

Pulse receivers, such as radar and pulse communications units, often use the tangential sensitivity as the measure of performance, which is the amplitude of pulse signal required to raise the noise level by its own RMS amplitude (Figure 16.15).

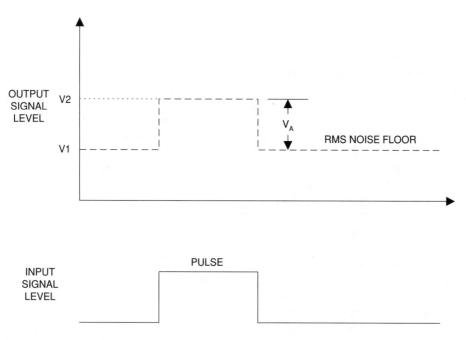

Fig. 16.15 *Tangential sensitivity.*

SELECTIVITY

Although no receiver specification is unimportant, if one had to choose between sensitivity and selectivity, the proper choice most of the time would be to take selectivity.

Selectivity is the measure of a receiver's ability to reject adjacent channel interference. Put another way, it's the ability to reject interference from signals on frequencies close to the desired signal frequency.

In order to understand selectivity requirements, one must first understand a little bit of the nature of radio signals. An unmodulated radio carrier theoretically has an infinitesimal (near-zero) bandwidth (although all real unmodulated carriers have a very narrow, but nonzero, bandwidth because they are modulated by noise and other artifacts). As soon as the radio signal is modulated to carry information, however, the bandwidth spreads. Even an on/off telegraphy (CW) or pulse signal spreads out either side of the carrier frequency an amount that is dependent on the sending speed and the shape of the keying waveform.

An AM signal spreads out an amount equal to twice the highest audio modulating frequencies. For example, a communications AM transmitter will have audio components from 300 to 3,000 Hz, so the AM waveform will occupy a spectrum that is equal to the carrier frequency (F) plus/minus the audio bandwidth ($F \pm 3,000$ Hz in the case cited). An FM carrier spreads out according to the *deviation*. For example, a narrowband FM landmobile transmitter with 5-kHz deviation spreads out ±5 kHz, while FM broadcast transmitters spread out ±75 kHz.

An implication of the fact that radio signals have bandwidth is that the receiver must have sufficient bandwidth to recover all of the signal. Otherwise, information may be lost and the output is distorted. On the other hand, allowing too much bandwidth increases the noise picked up by the receiver and thereby deteriorates the SNR. The goal of the selectivity system of the receiver is to match the bandwidth of the receiver to that of the signal. That is why receivers will use 270- or 500-Hz bandwidth for CW, 2 to 3 kHz for SSB, and 4 to 6 kHz for AM signals. They allow matching the receiver bandwidth to the transmission type.

The selectivity of a receiver has a number of aspects that must be considered: front-end bandwidth, IF bandwidth, IF shape factor, and the ultimate (distant frequency) rejection.

Front-End Bandwidth

The "front end" of a modern superheterodyne radio receiver is the circuitry between the antenna input terminal and the output of the first mixer stage. The reason why front-end selectivity is important is to keep out-of-band signals from afflicting the receiver. Transmitters located nearby can easily overload a poorly designed receiver. Even if these signals are not heard by the operator, they can desensitize a receiver, or create harmonics and intermodulation products that show up as "birdies" or other types of interference on the receiver. Strong local signals can take up a lot of the receiver's dynamic range, and thereby make it harder to hear weak signals.

In some crystal video microwave receivers, that front end might be wide open without any selectivity at all, but in nearly all other receivers there will be some form of frequency selection present.

Two forms of frequency selection are typically found. A designer may choose to use only one of them in a design. Alternatively, both might be used in the design, but separately (operator selection). Or finally, both might be used together. These forms can be called the resonant frequency filter (Figure 16.16A) and bandpass filter (Figure 16.16B) approaches.

The resonant frequency approach uses L-C elements tuned to the desired frequency to select which RF signals reach the mixer. In some receivers, these L-C elements are designed to track with the local oscillator that sets the operating frequency. That's why you see two-section variable capacitors for AM broadcast receivers with two different capacitance ranges for the two sections. One section tunes the LO and the other section tunes

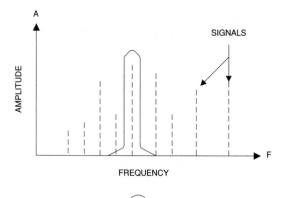

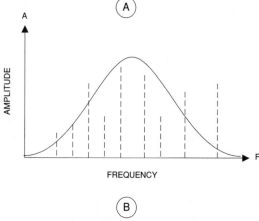

Fig. 16.16 *Two different bandwidths: (A) narrow and (B) broad.*

the tracking RF input. In other designs, a separate tuning knob ("preselector" or "antenna") is used.

The other approach uses a suboctave bandpass filter to admit only a portion of the RF spectrum into the front end. For example, a shortwave receiver that is designed to take the HF spectrum in 1-MHz pieces may have an array of RF input bandpass filters that are each 1 MHz wide (e.g., 9 to 10 MHz).

In addition to the reasons cited above, front-end selectivity also helps improve a receiver's image rejection and first IF rejection capabilities.

Image Rejection

An *image* in a superheterodyne receiver is a signal that appears at twice the IF distance from the desired RF signal and located on the opposite side of the LO frequency from the desired RF signal. In Figure 16.17, a superheterodyne operates with a 455-kHz (i.e., 0.455-MHz) IF and is turned to 24.0 MHz (F_{RF}). Because this receiver uses low-side LO injection, the LO frequency F_{LO} is 24.0 − 0.455 or 23.545 MHz. If a signal appears at twice the IF below the RF (i.e., 910 kHz below F_{RF}) and reaches the mixer, then it, too, has a difference frequency of 455 kHz, so it will pass right through the IF filtering as a valid signal. The image rejection specification tells how well this image frequency is suppressed. Normally, anything over about 70 dB is considered good.

Tactics to reduce image response vary with the design of the receiver. The best ap-

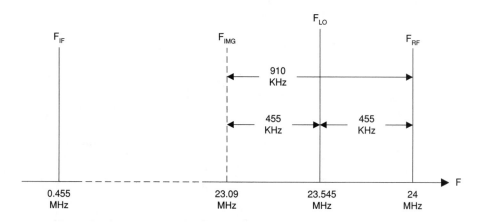

Fig. 16.17 *Image frequency.*

proach, at design time, is to select an IF frequency that is high enough that the image frequency will fall outside the passband of the receiver front end. Some HF receivers use an IF of 8.83 MHz, 9 MHz, 10.7 MHz, or something similar, and for image rejection these frequencies are considerably better than 455-kHz receivers in the higher HF bands. However, a common design trend is to do double conversion. In most such designs, the first IF frequency is considerably higher than the RF, being in the range 35 to 60 MHz (50 MHz is common in HF receivers, 70 MHz in microwave receivers).

The high IF makes it possible to suppress the VHF images with a simple low-pass filter. If the 24.0-MHz signal (above) were first up-converted to 50 MHz (74 MHz LO), for example, the image would be at 124 MHz. The second conversion brings the IF down to one of the frequencies mentioned above, or even 455 kHz. The lower frequencies are preferable to 50 MHz for bandwidth selectivity reasons because good-quality crystal, ceramic, or mechanical filters in the lower frequency ranges are easily available.

First IF Rejection

The first IF rejection specification refers to how well a receiver rejects radio signals op-

erating on the receiver's first IF frequency. For example, if your receiver has a first IF of 50 MHz, it must be able to reject radio signals operating on that frequency when the receiver is tuned to a different frequency. Although the shielding of the receiver is also an issue with respect to this performance, the front-end selectivity affects how well the receiver performs against first IF signals.

If there is no front-end selectivity to discriminate against signals at the IF frequency, then they arrive at the input of the mixer unimpeded. Depending on the design of the mixer, they then may pass directly through to the high-gain IF amplifiers and be heard in the receiver output.

IF Bandwidth

Most of the selectivity of the receiver is provided by the filtering in the IF amplifier section. The filtering might be L-C filters (especially if the principal IF is a low frequency such as 50 kHz), a ceramic resonator, a crystal filter, or a mechanical filter. Of these, the mechanical filter is usually regarded as best for narrow bandwidths, with the crystal filter and ceramic filters coming in next.

The IF bandwidth is expressed in kilohertz and is measured from the points on the IF frequency response curve where gain drops off −3 dB from the midband value

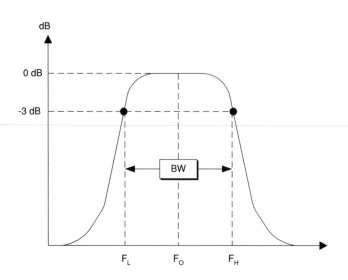

Fig. 16.18
IF bandwidth defined.

(Figure 16.18). This is why you will sometimes see selectivity specified in terms such as "6 kHz between –3 dB points."

The IF bandwidth must be matched to the bandwidth of the received signal for best performance. If too wide a bandwidth is selected, then the received signal will be noisy, and SNR deteriorates. If too narrow, then you might experience difficulties recovering all of the information that was transmitted. For example, an AM broadcast band radio signal has audio components out to 5 kHz, so the signal occupies up to 10 kHz of spectrum space ($F \pm 5$ kHz). If a 2.8-kHz SSB IF filter is selected, then it will tend to sound "mushy" and distorted.

IF Passband Shape Factor

The shape factor is a measure of the steepness of the receiver's IF passband and is taken by measuring the ratio of the bandwidth at –6 dB to the bandwidth at –60 dB (Figure 16.19A). The general rule is that the closer these numbers are to each other, the better the receiver. Anything in the 1:1.5 to 1:1.9 region can be considered high quality, while anything worse than 1:3 is not worth looking at for "serious" receiver uses. If the numbers are between 1:1.9 and 1:3, then the receiver could be regarded as being middling, but useful.

The importance of shape factor is that it modifies the notion of bandwidth. The cited bandwidth (e.g., 2.8 kHz for SSB) does not take into account the effects of strong signals that are just beyond those limits. Such signals can easily "punch through" the IF selectivity if the IF passband "skirts" are not steep. After all, the steeper they are, the closer a strong signal can be without messing up the receiver's operation. The situation is illustrated in Figure 16.19B. This curve inverts Figure 16.19A by plotting attenuation vs frequency. Assume that equal amplitude signals close to f_o are received (Figure 16.19C); the relative postfiltering amplitudes will match Figure 16.19D. Thus, selecting a receiver with a shape factor as close to the 1:1 ideal as possible will result in a more usable radio.

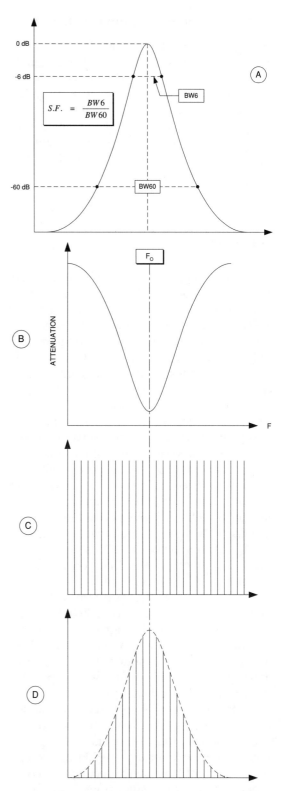

Fig. 16.19 *(A) Shape factor defined; (B–D) effect of shape factor on signal.*

Distant Frequency ("Ultimate") Rejection

This specification tells something about the receiver's ability to reject very strong signals that are located well outside the receiver's IF passband. This number is stated in negative decibels, and the higher the magnitude of the number, the better. An excellent receiver will have values in the –60 to –90 dB range, a middling receiver will see numbers in the –45 to –60 dB range, and a terrible receiver will be –44 dB or worse.

STABILITY

The stability specification measures how much the receiver frequency drifts as time elapses or temperature changes. The LO drift sets the overall stability of the receiver. This specification is usually given in terms of short-term drift and long-term drift (e.g., from LO crystal aging). The short-term drift is important in daily operation, while the long-term drift ultimately affects general dial calibration.

If the receiver is VFO controlled, or uses partial frequency synthesis (which combines VFO with crystal oscillators), then the stability is dominated by the VFO stability. In fully synthesized receivers, the stability is governed by the master reference crystal oscillator. If either an oven-controlled crystal oscillator (OCXO) or a temperature-compensated crystal oscillator (TCXO) is used for the master reference, then stability on the order of 1 part in 10^8/°C is achievable.

For most users, the short-term stability is what is most important, especially when tuning SSB, ECSS, or RTTY signals. A common specification value for a good receiver will be 50 Hz/hour after a 3-hour warm-up, or 100 Hz/hour after a 15-minute warm-up. The smaller the drift, the better the receiver.

The foundation of good stability is at design time. The local oscillator, or VFO portion of a synthesizer, must be operated in a cool, temperature-stable location within the equipment, and must have the correct type of components. Capacitor temperature coefficients are selected in order to cancel out temperature related drift in inductance values.

Post-design-time changes can also help, but these are less likely to be possible today than in the past. The chief cause of drift problems is heat. In the days of vacuum tube (valve) oscillators, the internal heating filament produced lots of heat that in turn created drift.

A related phenomenon seen on low-cost receivers, or certain homebrew receivers of doubtful merit, is mechanical frequency shifts. Although not seen on most modern receivers (even some very cheap designs), it was once a serious problem on less costly models. This problem is usually seen on VFO-controlled receivers in which vibration to the receiver cabinet imparts movement to either the inductor (L) or capacitor (C) element in an L-C VFO. Mechanically stabilizing these components will work wonders.

AGC RANGE AND THRESHOLD

Modern communications receivers must be able to handle signal strengths over a dynamic range of about 1,000,000:1. Tuning across a band occupied by signals of wildly varying strengths is hard on the ears and hard on the receiver's performance. As a result, most modern receivers have an automatic gain control (AGC) circuit that smoothes out these changes. The AGC will reduce gain for strong signals and increase it for weak signals (AGC can be turned off on most HF communications receivers). The AGC range is the change of input signal (in dBμV) from some reference level (e.g., 1 μV_{EMF}) to the input level that produces a 2-dB change in output level. Ranges of 90 to 110 dB are commonly seen.

The AGC threshold is the signal level at which the AGC begins to operate. If this threshold is set too low, the receiver gain will respond to noise and irritate the user. If it is set too high, the user will experience irritating shifts of output level as the band is tuned. AGC thresholds of 0.7 to 2.5 μV are common on decent receivers, with the better receivers being in the 0.7- to 1-μV range.

Another AGC specification sometimes seen deals with the speed of the AGC. Although sometimes specified in milliseconds, it is also frequently specified in subjective terms such as "fast" and "slow." This specification refers to how fast the AGC responds to changes in signal strength. If it is set too fast, then rapidly keyed signals (e.g., on–off CW) or noise transients will cause unnerving large shifts in receiver gain. If it is set too slow, then the receiver might as well not have an AGC. Many receivers provide two or more selections in order to accommodate different types of signals.

DYNAMIC PERFORMANCE

The dynamic performance specifications of a radio receiver are those which deal with how the receiver performs in the presence of very strong signals either cochannel or adjacent channel. Until about the 1960s, dynamic performance was somewhat less important than static performance for most users. However, today the role of dynamic performance is probably more critical than static performance because of crowded band conditions.

There are at least two reasons for this change in outlook. First, in the 1960s receiver designs evolved from tubes to solid-state. The new solid-state amplifiers were somewhat easier to drive into nonlinearity than tube designs. Second, there has been a tremendous increase in radio frequency signals on the air. There are far more transmitting stations than ever before, and there are far more sources of electromagnetic interference (EMI, pollution of the air waves) than in prior decades. With the advent of new and expanded wireless services available to an ever-widening market, the situation can only worsen. For this reason, it is now necessary to pay more attention to the dynamic performance of receivers than in the past.

INTERMODULATION PRODUCTS

Understanding the dynamic performance of the receiver requires knowledge of intermodulation products (IPs) and how they affect receiver operation. Whenever two signals are mixed together in a nonlinear circuit, a number of products are created according to the $mF_1 \pm nF_2$ rule, where m and n are either integers or zero (0, 1, 2, 3, 4, 5, . . .). Mixing can occur either in the mixer stage of a receiver front end or in the RF amplifier (or any outboard preamplifiers used ahead of the receiver) if the RF amplifier is overdriven by a strong signal.

It is also theoretically possible for corrosion on antenna connections, or even rusted antenna screw terminals to create IPs under certain circumstances. One even hears of cases where a rusty downspout on a house rain gutter caused reradiated mixed signals.

The spurious IP signals are shown graphically in Figure 16.20. The order of the product is given by the sum $(m + n)$. Given input signal frequencies of F_1 and F_2, the main IPs are:

Second-order:	$F_1 \pm F_2$
	$2F_1$
	$2F_2$
Third-order:	$2F_1 \pm F_2$
	$2F_2 \pm F_1$
	$3F_1$
	$3F_2$
Fifth-order:	$3F_1 \pm 2F_2$
	$3F_2 \pm 2F_1$
	$5F_1$
	$5F_2$

When an amplifier or receiver is overdriven, the second-order content of the output signal increases as the square of the input signal level, while the third-order responses increase as the cube of the input signal level.

Consider the case where two HF signals, $F_1 = 10$ MHz and $F_2 = 15$ MHz, are mixed together. The second-order IPs are 5 and 25 MHz; the third-order IPs are 5, 20, 35 and 40 MHz; and the fifth-order IPs are 0, 25,

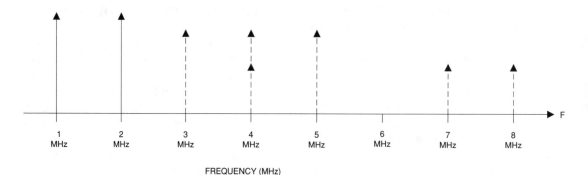

FREQUENCY (MHz)

Fig. 16.20 *IMD spectrum.*

60, and 65 MHz. If any of these are inside the passband of the receiver, then they can cause problems. One such problem is the emergence of "phantom" signals at the IP frequencies. This effect is seen often when two strong signals (F_1 and F_2) exist and can affect the front end of the receiver, and one of the IPs falls close to a desired signal frequency, F_d. If the receiver were tuned to 5 MHz, for example, a spurious signal would be found from the F_1–F_2 pair given above.

Another example is seen from strong in-band, adjacent-channel signals. Consider a case where the receiver is tuned to a station at 9,610 kHz, and there are also very strong signals at 9,600 kHz and 9,605 kHz. The near (in-band) IP products are:

3rd-order: 9,595 kHz (ΔF = 15 kHz)
9,610 kHz (ΔF = 0 kHz)
(on channel!)

5th-order: 9,590 kHz (ΔF = 20 kHz)
9,615 kHz (ΔF = 5 kHz).

Note that one third-order product is on the same frequency as the desired signal, and could easily cause interference if the amplitude is sufficiently high. Other third- and fifth-order products may be within the range where interference could occur, especially on receivers with wide bandwidths.

The IP orders are theoretically infinite because there are no bounds on either m or n. However, in practical terms, because each successively higher order IP is reduced in ampli-

tude compared with its next lower order mate, only the second-order, third-order, and fifth-order products usually assume any importance. Indeed, only the third-order is normally used in receiver specifications sheets because they fall close to the RF signal frequency.

There are a large number of IMD products from just two signals applied to a nonlinear medium. But consider the fact that the two-tone case used for textbook discussions is rarely encountered in actuality. A typical two-way radio installation is in a signal-rich environment, so when dozens of signals are present the number of possible combinations climbs to an unmanageable extent.

–1 dB COMPRESSION POINT

An amplifier produces an output signal that has a higher amplitude than the input signal. The transfer function of the amplifier (indeed, any circuit with output and input) is the ratio OUT/IN, so for the power amplification of a receiver RF amplifier it is P_o/P_{in} (or, in terms of voltage, V_o/V_{in}). Any real amplifier will saturate given a strong enough input signal (see Figure 16.21). The dotted line represents the theoretical output level for all values of input signal (the slope of the line represents the gain of the amplifier). As the amplifier saturates (solid line), however, the actual gain begins to depart from the theoretical at some level of input signal (P_{in1}). The –1 dB compression point is that output level

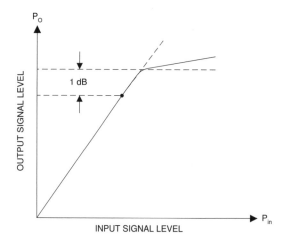

Fig. 16.21 *-1 dB compression point.*

at which the actual gain departs from the theoretical gain by –1 dB.

The –1 dB compression point is important when considering either the RF amplifier ahead of the mixer (if any) or any outboard preamplifiers that are used. The –1 dB compression point is the point at which intermodulation products begin to emerge as a serious problem. It is also the case that harmonics are generated when an amplifier goes into compression. A sine wave is a "pure" signal because it has no harmonics (all other wave shapes have a fundamental plus harmonic frequencies). When a sine wave is distorted, however, harmonics arise. The effect of the compression phenomenon is to distort the signal by clipping the peaks, thus raising the harmonics and intermodulation distortion products.

THIRD-ORDER INTERCEPT POINT

It can be claimed that the third-order intercept point (TOIP) is the single most important specification of a receiver's dynamic performance because it predicts the performance as regards intermodulation, cross-modulation, and blocking desensitization.

Third-order (and higher) intermodulation products (IPs) are normally very weak

and don't exceed the receiver noise floor when the receiver is operating in the linear region. But as input signal levels increase, forcing the front end of the receiver toward the saturated nonlinear region, the IPs emerge from the noise (Figure 16.22) and begins to cause problems. When this happens, new spurious signals appear on the band and self-generated interference begins to arise.

Figure 16.23 shows a plot of the output signal vs fundamental input signal. Note the output compression effect that was seen earlier in Figure 16.20. The dotted gain line continuing above the saturation region shows the theoretical output that would be produced if the gain did not clip. It is the nature of third-order products in the output signal to emerge from the noise at a certain input level and increase as the cube of the input level. Thus, the slope of the third-order line increases 3 dB for every 1-dB increase in the response to the fundamental signal. Although the output response of the third-order line saturates similarly to that of the fundamental signal, the gain line can be continued to a point where it intersects the gain line of the fundamental signal. This point is the third-order intercept point.

Interestingly enough, one receiver feature that can help reduce IP levels back down under the noise is the use of a front-end attenuator (aka *input attenuator*). In the presence of strong signals even a few decibels of input attenuation is often enough to drop the IPs back into the noise, while afflicting the desired signals by only a small amount.

Other effects that reduce the overload caused by a strong signal also help. Situations arise where the apparent third-order performance of a receiver improves dramatically when a lower gain antenna is used. This effect can be easily demonstrated using a spectrum analyzer for the receiver. This instrument is a swept frequency receiver that displays an output on an oscilloscope screen that is amplitude vs frequency, so a single signal shows as a spike. In one test, a strong, local VHF band repeater came on the air every few seconds,

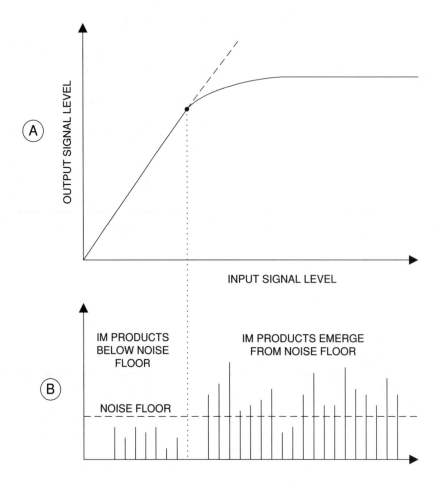

Fig. 16.22 *Signal above –1 dB compression point will generate IMD noise.*

and one could observe the second- and third-order IPs along with the fundamental repeater signal. There were also other strong signals on the air, but just outside the band. Inserting a 6-dB barrel attenuator in the input ("antenna") line eliminated the IP products, showing just the actual signals. Rotating a directional antenna away from the direction of the interfering signal will also have this effect in many cases.

Preamplifiers are popular receiver accessories, but can often reduce rather than enhance performance. Two problems commonly occur (assuming the preamp is a low-noise device). The best-known problem is that the preamp amplifies noise as much as signals, and while it makes the signal louder,

it also makes the noise louder by the same amount. Since it's the signal-to-noise ratio that is important, one does not improve the situation. Indeed, if the preamp is itself noisy, it will deteriorate the SNR. The other problem is less well known, but potentially more devastating. If the increased signal levels applied to the receiver drive the receiver nonlinear, then IPs begin to emerge.

When evaluating receivers, a TOIP of +5 to +20 dBm is excellent performance, while up to +27 dBm is relatively easily achievable, and +35 dBm has been achieved with good design; anything greater than +50 dBm is close to miraculous (but attainable). Receivers are still regarded as good performers in the 0 to +5 dBm range, and mid-

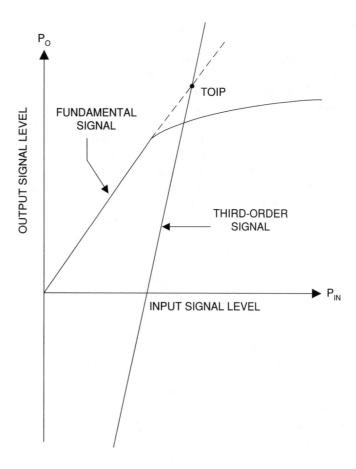

Fig. 16.23
Third-order intercept point defined.

dling performers in the −10 to 0 dBm range. Anything below −10 dBm is not usually acceptable. A general rule is to buy the best third-order intercept performance that you can afford, especially if there are strong signal sources in your vicinity.

DYNAMIC RANGE

The dynamic range of a radio receiver is the range (measured in decibels) from the minimum discernible signal to the maximum allowable signal. Although this simplistic definition is conceptually easy to understand, in the concrete it's a little more complex. Several definitions of dynamic range are used.

One definition of dynamic range is that it is the input signal difference between the sensitivity figure (e.g., 0.5 μV for 10 dB $S+N/N$) and the level that drives the receiver far enough into saturation to create a certain amount of distortion in the output. This definition was common on consumer broadcast band receivers at one time (especially automobile radios, where dynamic range was somewhat more important due to mobility). A related definition takes the range as the distance in decibels from the sensitivity level and the −1 dB compression point. In still another definition, the *blocking dynamic range* is the range of signals from the sensitivity level to the blocking level (see below).

A problem with these definitions is that they represent single-signal cases and so do not address the receiver's dynamic characteristics. There is both a "loose" and a more formal definition that is somewhat more useful, and is at least standardized. The loose version is that dynamic range is the range of signals over which dynamic effects (e.g., intermodulation) do not exceed the noise floor of the receiver. For HF receivers the recommended dynamic range is usually two-thirds the difference between the noise floor and the third-order intercept point in a 3-kHz bandwidth.

There is also an alternative definition: dynamic range is the difference between the fundamental response input signal level and the third-order intercept point along the noise floor, measured with a 3-kHz bandwidth. For practical reasons, this measurement is sometimes made not at the actual noise floor (which is sometimes hard to ascertain), but rather at 3 dB above the noise floor.

A certain measurement procedure produces similar results (the same method is used for many amateur radio magazine product reviews). Two equal-strength signals are input to the receiver at the same time. The frequency difference has traditionally been 20 kHz for HF and 30 to 50 kHz for VHF receivers (modern band crowding may indicate a need for a specification at 5-kHz separation on HF). The amplitudes of these signals are raised until the third-order distortion products are raised to the noise floor level. For 20-kHz spacing, using the two-signal approach, anything over 90 dB is an excellent receiver, while anything over 80 dB is at least decent.

The difference between the single-signal and two-signal (dynamic) performance is not merely an academic exercise. Besides the fact that the same receiver can show as much as 40 dB difference between the two measures (favoring the single-signal measurement), the most severe effects of poor dynamic range show up most in the dynamic performance.

BLOCKING

The blocking specification refers to the ability of the receiver to withstand very strong off-tune signals that are at least 20 kHz away from the desired signal, although some use 100-kHz separation. When very strong signals appear at the input terminals of a receiver, they may desensitize the receiver, i.e.,

reduce the apparent strength of desired signals over what they would be if the interfering signal were not present.

Figure 16.24 shows the blocking behavior. When a strong signal is present, it takes up more of the receiver's resources than normal, so there is not enough of the output power budget to accommodate the weaker desired signals. But if the strong undesired signal is turned off, then the weaker signals receive a full measure of the unit's power budget.

The usual way to measure blocking behavior is to input two signals, a desired signal at 60 dBμV and another signal 20 (or 100) kHz away at a much stronger level. The strong signal is increased to the point where blocking desensitization causes a 3-dB drop in the output level of the desired signal. A good receiver will show ≥90 dBμV, with many being considerably better. An interesting note about modern receivers is that the blocking performance is so good that it's often necessary to specify the input level difference (dB) that causes a 1-dB drop, rather than a 3-dB drop, of the desired signal's amplitude.

The phenomenon of blocking leads us to an effect that is often seen as paradoxical on first blush. Many receivers are equipped with front-end attenuators that permit fixed attenuation values of 1, 3, 6, 12, or 20 dB (or some subset) to be inserted into the signal path ahead of the active stages. When a strong signal that is capable of causing desensitization is present, adding attenuation often increases the level of the desired signals in the output, even though overall gain is reduced. This occurs because the overall signal that the receiver front end is asked to handle is below the threshold where desensitization occurs.

CROSS-MODULATION

Cross-modulation is an effect in which amplitude modulation (AM) from a strong undesired signal is transferred to a weaker desired signal. Testing is usually done (in HF receivers) with a 20-kHz spacing between the

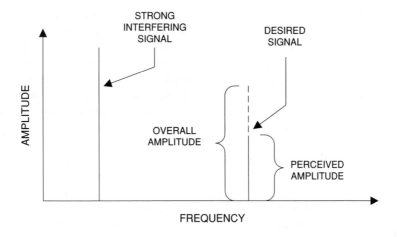

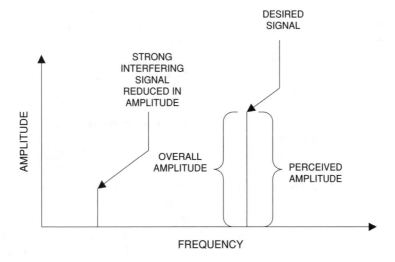

Fig. 16.24 *Perceived amplitude of signal.*

desired and undesired signals, a 3-kHz IF bandwidth on the receiver, and the desired signal set to 1,000 μV_{EMF} (–53 dBm). The undesired signal (20 kHz away) is amplitude modulated to the 30 percent level. This undesired AM signal is increased in strength until an unwanted AM output 20 dB below the desired signal is produced.

A cross-modulation specification ≥ 100 dB would be considered decent performance. This figure is often not given for modern HF receivers, but if the receiver has a good third-order intercept point, it is likely to also have good cross-modulation performance.

Cross-modulation is also said to occur naturally, especially in transpolar and North Atlantic radio paths where the effects of the aurora are strong. According to one (possibly apocryphal) legend, there was something called the "Radio Luxembourg Effect" discovered in the 1930s. Modulation from a very strong broadcaster (BBC) appeared on the Radio Luxembourg signal received in North America. This effect was said to be an ionospheric cross-modulation phenomenon and apparently occurs when the strong station is within 175 miles of the great circle path between the desired station and the receiver site.

RECIPROCAL MIXING

Reciprocal mixing occurs when noise side-bands from the local oscillator (LO) signal in a superheterodyne receiver mix with a strong undesired signal that is close to the desired signal. Every oscillator signal produces noise, and that noise tends to amplitude modulate the oscillator's output signal. It will thus form sidebands either side of the LO signal. The production of phase noise in all LOs is well known, but in more recent designs the digitally produced synthesized LOs are prone to additional noise elements as well. The noise is usually measured in –dBc (decibels below carrier, or, in this case, decibels below the LO output level).

In a superheterodyne receiver, the LO beats with the desired signal to produce an intermediate frequency (IF) equal to either the sum (LO + RF) or difference (LO – RF). If a strong unwanted signal is present, then it might mix with the noise sidebands of the LO to reproduce the noise spectrum at the IF frequency (see Figure 16.25). In the usual test scenario, the reciprocal mixing is defined as the level of the unwanted signal (dB) at 20 kHz required to produce a noise sideband 20 dB down from the desired IF signal in a specified bandwidth (usually 3 kHz on HF receivers). Figures of –90 dBc or better are considered good.

The importance of the reciprocal mixing specification is that it can seriously deteriorate the observed selectivity of the receiver, yet is not detected in the normal static measurements made of selectivity (it is a "dynamic selectivity" problem). When the LO noise sidebands appear in the IF, the distant frequency attenuation (>20 kHz off-center of a 3-kHz bandwidth filter) can deteriorate 20 to 40 dB.

The reciprocal mixing performance of receivers can be improved by eliminating the noise from the oscillator signal. Although this sounds simple, in practice it is often quite difficult. A tactic that works well is to add high-Q filtering between the LO output and the mixer input. The narrow bandwidth of the high-Q filter prevents excessive noise side-bands from getting to the mixer. Although this sounds like quite the easy solution, as they say, "The devil is in the details."

IF NOTCH REJECTION

If two signals fall within the passband of a receiver, they will both compete to be heard. They will also heterodyne together in the detector stage, producing an audio tone equal to their carrier frequency difference. For example, suppose we have an AM receiver with a 5-kHz bandwidth and a 455-kHz IF. If

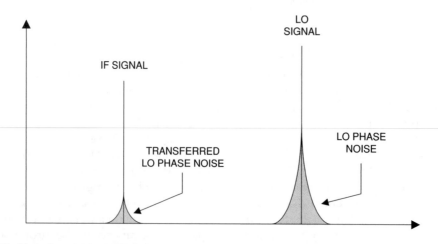

Fig. 16.25   Transferred LO phase noise.

two signals appear on the band such that one appears at an IF of 456 kHz and the other is at 454 kHz, then both are within the receiver passband and both will be heard in the output. However, the 2-kHz difference in their carrier frequency will produce a 2-kHz heterodyne audio tone difference signal in the output of the AM detector.

In some receivers, a tunable, high-Q (narrow and deep) notch filter is in the IF amplifier circuit. This tunable filter can be turned on and then adjusted to attenuate the unwanted interfering signal, reducing the irritating heterodyne. Attenuation figures for good receivers vary from –35 to –65 dB or so (the more negative the better).

There are some trade-offs in notch filter design. First, the notch filter Q is more easily achieved at low IF frequencies (such as 50 to 500 kHz) than at high IF frequencies (e.g., 9 MHz and up). Also, the higher the Q, the better the attenuation of the undesired squeal, but the touchier it is to tune. Some happy middle ground between the irritating squeal and the touchy tune is mandated here.

Some receivers use audio filters rather than IF filters to help reduce the heterodyne squeal. In the AM broadcast band, channel spacing is typically 8 to 10 kHz (depending on the part of the world), and the transmitted audio bandwidth (hence the sidebands) are 5 kHz. Designers of AM BCB receivers usually insert an R-C low-pass filter with a –3 dB point just above 4 or 5 kHz right after the detector in order to suppress the audio heterodyne. This R-C filter is called a "tweet filter" in the slang of the electronic service/repair trade.

Another audio approach is to sharply limit the bandpass of the audio amplifiers.

For AM BCB reception, a 5-kHz bandpass is sufficient, so the frequencies higher can be rolled off at a fast rate in order to produce only a small response an octave higher (10 kHz). In shortwave receivers, this option is weaker because the station channels are typically 5 kHz, and many don't bother to honor the official channels anyway. And on the amateur radio bands, frequency selection is a perpetually changing ad-hocracy, at best. Although the shortwave bands typically only need 3-kHz bandwidth for communications and 5-kHz for broadcast, the tweet filter and audio roll-off might not be sufficient. In receivers that lack an effective IF notch filter, an audio notch filter can be provided.

INTERNAL SPURII

All receivers produce a number of internal spurious signals that sometimes interfere with the operation. Both old and modern receivers have spurious signals from assorted high-order mixer products, from power-supply harmonics, parasitic oscillations, and a host of other sources. Newer receivers with either (or both) synthesized local oscillators and digital frequency readouts produce noise and spurious signals in abundance (note: low-power digital chips with slower rise times—CMOS, NMOS, etc.—are generally much cleaner than higher power, fast rise-time chips such as TTL devices). With appropriate filtering and shielding, it is possible to hold the "spurs" down to –100 dB relative to the main maximum signal output, or within about 3 dB or the noise floor, whichever is lower.

Chapter 17

Dealing with Radio Receiver System EMI

There are several strategies that can improve the EMI and noise performance of a radio communications receiver. Some methods are available for after-market use with existing receivers, while others are only practical when designing a new receiver. Before studying this chapter you should be familiar with receiver technology (see Chapter 16). Additional information on solving receiver EMI is found in Chapter 12 dealing with interference to television receivers. You are advised to read that chapter also.

One method for preventing the dynamic effects of EMI is to reduce the signal level applied to the input of the receiver at all frequencies. A front-end attenuator will help here. Indeed, many modern receivers have a switchable attenuator built into the front-end circuitry. This attenuator will reduce the level of all signals, backing the receiver circuits away from the compression point and, in the process, eliminating many of the dynamic problems described above.

Another method is to prevent the offending signal or signals from ever reaching the receiver front end. This goal can be achieved by using any of several bandpass filters, high-pass filters, low-pass filters, or tunable filters (as appropriate) ahead of the receiver front end. These filters may not help if the intermodulation products fall close to the desired frequency (e.g., the third-order difference products $2F_1 - F_2$ and $2F_2 - F_1$), but for other frequencies filters are quite useful.

When designing a receiver there are some things that can be done. If an RF amplifier is used between the antenna and mixer, then use either field effect transistors (MOSFETs are popular) or a relatively high-powered but low-noise bipolar (NPN or PNP) transistor intended for cable-TV applications, such as the 2N5109 or MRF-586. Gain in the RF amplifier should be kept minimal (4 to 8 dB). It is also useful to design the RF amplifier with as high a DC power supply potential as possible. That tactic allows a lot of "headroom" as input signals become larger. Slew-rate symmetry is an important characteristic to have in an RF amplifier. This means that the rise time and fall time of the amplifier are equal, or nearly so.

You might also wish to consider deleting the RF amplifier altogether. One receiver design philosophy holds that one should use a high-dynamic-range mixer with only input tuning or suboctave filters between the antenna connector and the mixer's RF input. It is possible to achieve noise figures of 10 dB or so using this approach, and that is sufficient for HF work (although it may be marginal for VHF and up). The tube-design 1960s-vintage Squires-Sanders SS-1 receiver used this approach and delivered superior performance. The front-end mixer valve was the 7360 double-balanced mixer.

The mixer in a newly designed receiver project should be a high-dynamic-range type regardless of whether an RF amplifier is used. Popular with some designers is the double-balanced switching type mixer. Although examples can be fabricated from MOSFET transistors and MOS digital switches, there are also some integrated circuit versions on the market that are intended as MOSFET switching mixers. Passive DBMs are available that operate up to RF levels of +20 dBm.

INTERMOD HILL: A TALE OF WOE

There is a hill not too far from the author where radio receivers are sorely tested. I first experienced this hill when a friend of mine got his ham radio operator's license and lived only a block from the site. This hilltop (actually, "bumptop" is more like it) has a 2,000-watt AM BCB station, two 50-kW FM BCB stations, and a microwave relay tower with scads of antennas on it. Not helping things much are the scores upon scores of two-way landmobile radio antennas. It seems that landmobile and cellular operators rent space on radio station towers to gain the height they need.

So what? Doesn't that just give you a signal-rich environment? No! The problem is that your receiver can only handle a certain amount of radio frequency energy in its front end. That area of the receiver is not very narrowband, so signals wander through that might not otherwise make it. Indeed, even if your front end is tuned (usually with an "antenna tune" or "preselector" control) the bandpass is quite broad. Receivers with an untuned front end may use a bandpass filter, and that is still a problem.

The offending signal need not be in the same band as the signal being sought. Indeed, the offending signal may be out of band and not even heard on the receiver. If it gets to the RF amplifier or mixer, then it may take up enough of the receiver's dynamic range to desensitize your receiver . . . making it less sensitive, in case you didn't get the point.

Another manifestation is that the offending signal may cause harmonics of itself to be generated. Those harmonics are accepted as valid signals, so the receiver will tune them in as if they arrived on the antenna. The problem is that the undesired signal drives the receiver front end into a nonlinear operating region, and that has the effect of increasing harmonic distortion.

Still another problem caused by letting strong out-of-band signals reach the receiver is the problem of intermodulation products. When only two signal frequencies are present and are able to drive the front-end circuitry of the receiver into nonlinearity, then a batch of new frequencies are generated. After all, the front end contains a mixer, and if the signals hit the radio-frequency amplifier first, it can make that stage pretend it's a mixer. The frequencies generated are defined by $mF_1 \pm nF_2$, where m and n are integers (0, 1, 2, 3, . . . N). These frequencies are:

$F_1 + 2F_2$	Third-order products
$F_1 - 2F_2$	
$2F_1 + F_2$	
$2F_1 - F_2$	
$2F_1 + 3F_2$	Fifth-order products
$2F_1 - 3F_2$	
$3F_1 + 2F_2$	
$3F_1 - 2F_2$	
etc.	

So what does this mean? Suppose you live near an AM broadcast band station that is operating on 1,500 kHz. Suppose a local ham operator starts broadcasting on a frequency of 7,200 kHz. These frequencies will provide third- and fifth-order intermodulation products of 4,200, 9,900, 10,200, 12,900, 15,900, 18,600, 18,900, and 24,600 kHz.

Note where those intermodulation products fall? Right in the middle of some of your listening territory. And when you count the fact that the ham operator can operate over a 300-kHz portion of the band from 7,000 to 7,300 kHz, there is actually a wide area 200 kHz below and 100 kHz above those spot frequencies that is vulnerable.

Keep in mind that AM and FM BCB stations tend to have two attributes that help them mess up each other's reception: (1) they are local, so their strength is very high at the receiver's location; and (2) they are high-powered (250 watts to 50 kilowatts). As a result, there is a huge signal from that BCB station present at the receiver. If the receiver antenna is on rented tower space (on a BCB tower), then signal levels are huge. Add to the BCB signals the dozens of microwave, two-way, repeater, and cellular signals present—then you can see how a simple equation such as $mF_1 \pm nF_2$ can explode into a huge number of signals if IMD occurs. Is Intermod Hill rare? No— it's common for radio operators to seek height for antennas so they tend to cluster.

THE PROBLEM

So what is the problem? Your receiver, no matter what frequency it receives, is designed to accept only a certain maximum amount of radio frequency energy in the front end. If more energy is present, then one or more of several overload conditions results. The overload could result from a desired station being tuned too strong. In other cases, there are simply too many signals within the passband for the receiver front end to accommodate. In still other cases, a strong out-of-band signal is present.

Figure 17.1 shows several conditions that your receiver might have to survive. Figure 17.1A is the ideal situation. Only one signal exists on the band, and it is centered in the passband of the receiver. This never happens, and the problem has existed since Marconi was hawking interest in his fledgling wireless company. Radio pioneers Fessenden and Marconi interfered with each other while reporting

yacht races off Long Island around the turn of the century. Not an auspicious beginning for maritime wireless! A more realistic situation is shown in Figure 17.1B. Here we see a large number of signals both in and out of the band signals, both weaker and stronger than the desired signal. Another situation is shown in Figure 17.1C where an extremely strong local station (e.g., AM BCB signal) is present, but is out of the receiver's front-end passband. The situation that you probably face is shown in Figure 17.1D: a large out-of-band AM BCB signal as well as the usual huge number of other signals both in and out of band.

Several different problems result from this situation, all of which are species of front-end overload-caused intermodulation and/or cross-modulation.

Blanketing

If you tune across the shortwave bands, especially those below 10 or 12 MHz, and note an AM BCB signal that seems like it is hundreds of kilohertz wide, then you are witnessing blanketing. It drives the mixer or RF amplifier of the receiver into severe nonlinearity, producing a huge number of spurious signals, and apparently a very wide bandwidth.

Desensitization

Your receiver can only accommodate a limited amount of RF energy in the front-end circuits. This level is expressed in the *dynamic range* specification of the receiver, and is hinted by the *third-order intercept point* (TOIP) and *−1 dB compression point* specifications. Figure 17.2A shows what happens in desensitization situations when a strong out-of-band signal is present. The strong out-of-band signal takes up so much of the dynamic range "headroom" that only a small amount of capacity remains for the desired signal. The level of the desired signal is thereby reduced. In some cases, the overload is so severe that the desired signal becomes inaudible. If you can filter out or otherwise attenuate the strong out-of-band signal (Figure 17.2B), then the headroom is restored, and the receiver has plenty of capacity to ac-

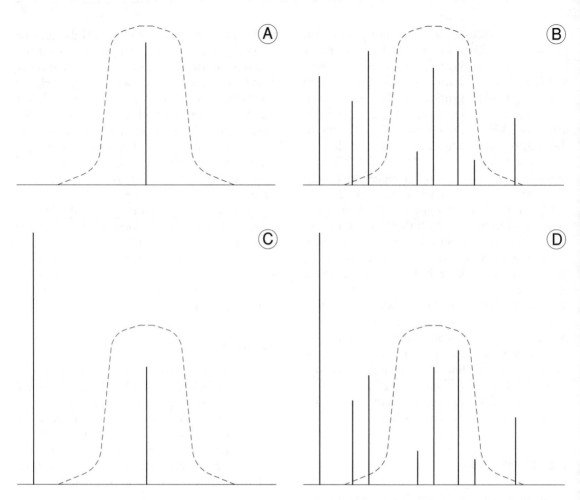

Fig. 17.1 *(A) Ideal situation; (B) more common situation; (C) situation where there is a strong off-channel signal; (D) unusual but fairly common situation.*

commodate both signals without causing other problems.

Figure 17.3 shows two more situations. In Figure 17.3A we see the response of the receiver when output level is plotted as a function of input signal level. The ideal situation is shown by the dotted line from the (0,0) intercept to an infinitely strong signal. Real radio receivers depart from the ideal and eventually saturate (solid line beyond the dot). The point denoted by the dot on the solid line is the point at which the TOIP is figured, but that's a topic covered in Chapter 16. What's important here is to consider what happens when signals are received that are stronger than the input signal that produces the flattening of the response.

In Figure 17.3B we see the generation of harmonics, i.e., integer (1, 2, 3, . . .) multiples of the offending signal's fundamental frequency. These harmonics may fall within the passband of your receiver and are seen as valid signals even though they were generated in the receiver itself!

The intermodulation problem is shown in Figure 17.3C. It occurs when two or more signals are present at the same time. The strong intermodulation products are created when two of these signals heterodyne together. The heterodyne ("mixing") action occurs because the receiver front end is nonlinear at this point. The frequencies produced by just two input frequencies (F1 and F2) are described by $mF_1 \pm nF_2$, where m

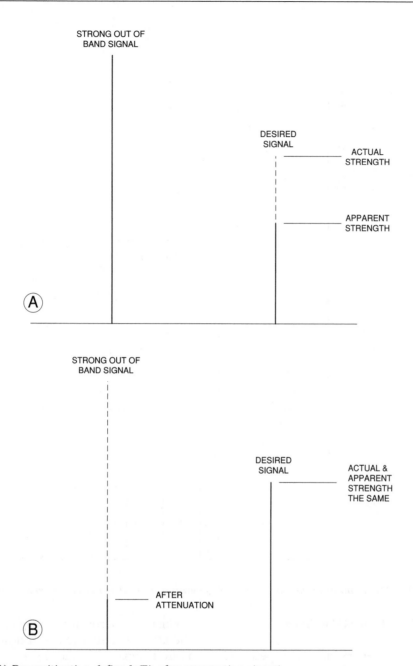

Fig. 17.2 *(A) Desensitization defined; (B) after-attenuation situation.*

and *n* are integers. As you can see, depending on how many frequencies are present and how strong they are, a huge number of spurious signals can be generated by the receiver front end.

What about IF selectivity? Suppose there is an IF filter of 2.7 to 8 kHz (depending on model and mode). So why doesn't it reject the dirty smelly bad-guy signals? The problem is that the damage occurs in the front-end section of the receiver, before the signals encounter the IF selectivity filters. The problem is due to an overdriven RF amplifier, mixer, or both. The only way to deal with this problem is to reduce the level of the offending signal.

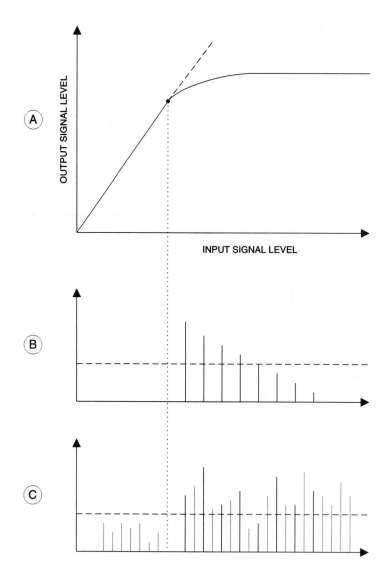

Fig. 17.3 *(A) Output-vs-input graph; (B) harmonic generation; (C) intermodulation distortion situation.*

THE ATTENUATOR SOLUTION

Some modern receivers are equipped with one or more switchable attenuators in the front end. Some receivers also include an RF gain control that sometimes operates in the same manner. Some receiver operators use external switchable attenuators for exactly the same purpose. The idea behind the attenuator is to reduce all of the signals to the front end enough to drop the overall energy in the circuit to below the level that can be accommodated without either overload or inter-

modulation occurring at significant levels. The attenuator reduces both desired and undesired signals, but the perceived ratio is altered when the receiver front end is deloaded to a point where desensitization occurs, or intermods and harmonics pop up.

THE ANTENNA SOLUTION

The antenna that you select can make some difference in EMI problems. Generally, a resonant antenna with its end nulls pointed toward the offending station will provide better performance than an omnidirectional antenna. Also, it is well known that vertical HF antennas are more susceptible to EMI because they respond better to the ground-wave electrical field generated by the station.

THE FILTER SOLUTION

One of the best EMI solutions is to filter out the offending signals before they hit the receiver front end in a manner that affects the desired signals only minimally. This task is not possible with the attenuator solution, which is an "equal opportunity" situation because it affects all signals equally. Figure 17.4A shows what happens to a signal that is outside the passband of a frequency-selective filter: it is severely attenuated. It does not drop to zero, but the reduction can be quite profound in some designs.

Signals within the receiver's passband are not unaffected by the filter, as shown in Figure 17.4B. The loss for in-band signals is, however, considerably less than for out-of-band signals.

This loss is called insertion loss and is usually quite small (1 or 2 dB) compared to the loss for out-of-band signals (lots of decibels).

Several different types of filter are used in reducing interference depending on circumstance. A high-pass filter passes all signals above a specified cutoff frequency (F_c). The low-pass filter passes all signals below the cutoff frequency. A bandpass filter passes signals between a lower cutoff frequency (F_L) and an upper cutoff frequency (F_H). A stopband filter is just the opposite of a bandpass filter: it stops signals on frequencies between F_L and F_H, while passing all others. A *notch* filter, also called a wavetrap, will stop a particular frequency (F_o), but not a wide band of frequencies as does the stop-band filter. In all cases, these filters stop the frequencies in the designated band, while passing all others.

The positioning of the filter in the antenna system is shown in Figure 17.5. The ideal location is as close as possible to the antenna input connector. The best practice, if space is available, is to use a double-male coaxial connector to connect filter output connector to the antenna input connector on the receiver. A

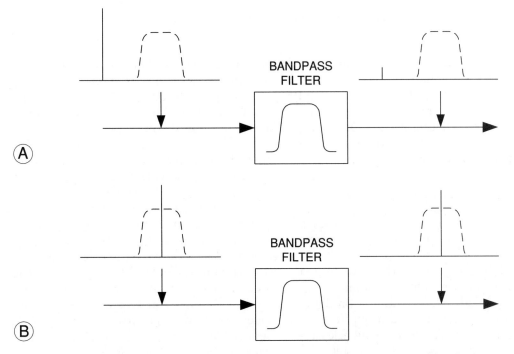

Fig. 17.4   *(A) Bandpass filter on out-of-band signal; (B) bandpass filter on in-band signal.*

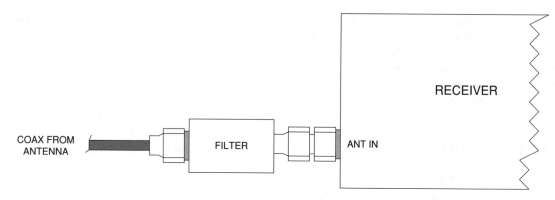

Fig. 17.5 *Placement of filter on receiver.*

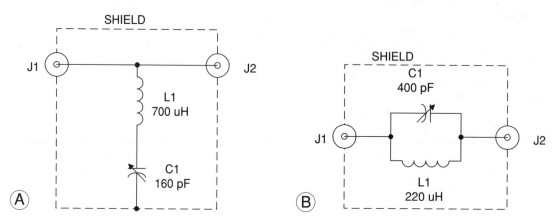

Fig. 17.6 *(A) Series-tuned wavetrap filter; (B) parallel-tuned wavetrap filter.*

short piece of coaxial cable can connect the two terminals if this approach is not suitable in your case. Be sure to earth both the ground terminal on the receiver and the ground terminal of the filter (if one is provided). Otherwise, depend on the coaxial connectors' outer shell making the ground connection.

Wavetraps

A wavetrap is a tuned circuit that causes a specific frequency to be rejected. Two forms are used: series tuned (Figure 17.6A) and parallel tuned (Figure 17.6B). The series-tuned version is placed across the signal line (as in Figure 17.6A) and works because it produces a very low impedance at its resonant frequency and a high impedance at frequencies removed from resonance. As a result, the interfering signal will see a resonant series-tuned wavetrap as a near short circuit, while other frequencies do not. The parallel resonant form is placed in series with the antenna line (as in Figure 17.6B). It provides a high impedance to its resonant frequency, so it will block the offending signal before it reaches the receiver. It provides a low impedance to frequencies removed from resonance.

The wavetraps are useful in situations where a single station is causing a problem and there is no desire to eliminate nearby stations—for example, if the receiver is close to a medium-wave AM BCB signal, and the operator doesn't want to interrupt reception of other AM BCB signals. The values of components shown in Figures 17.6A and 17.6B are suitable for the MW AM BCB, but can be scaled to the other bands if desired.

If there are two stations causing significant interference, then two wavetraps will have to be provided, separated by a short piece of coaxial cable, or contained within separate compartments of the same shielded box. In that case, use a parallel-tuned wavetrap for one frequency, and a series-tuned wavetrap for the other. Otherwise, interaction between the wavetraps can cause problems.

High-Pass Filters

One very significant solution to EMI from an AM BCB station is to use a high-pass filter with a cutoff frequency between 1,700 and 3,000 kHz. It will pass the shortwave frequencies and severely attenuate AM BCB signals, causing the desired improvement in performance. Figure 17.7 shows a design used for many decades. It is easily built because the capacitor values are 0.001 and 0.002 µF (which some people make by parallel connecting two 0.001-µF capacitors). The inductors are both 3.3 µH, so they can be made with toroid cores. If the T-50-2 RED cores are used (A_L = 49), then 26 turns of small diameter enameled wire will suffice. If the T-50-15 RED/WHITE cores are used (A_L = 135), then 15 turns are used. The circuit of Figure 17.7 produces fairly decent results for low effort.

Absorptive Filters

The absorptive filter solves a problem with the straight high-pass filter method and produces generally better results at the cost of more complexity. This filter (Figure 17.8) consists of

Table 17.1   Filter Capacitors Values

C1: 1,820 pF	Use 1000 pF (0.001 µF or 1 nF) in parallel with 820 pF
C2: 1,270 pF	Use 1000 pF and 270 pF in parallel
C6: 1,400 pF	Use 1000 pF, 180 pF and 220 pF in parallel

The other capacitors are standard values.

a high-pass filter (C4-C6/L4-L6) between the antenna input (J1) and the receiver output (J2). It passes signals above 3 MHz and rejects those below that cutoff frequency. It also has a low-pass filter (C1-C3/L1-L3) that passes signals below 3 MHz. What is notable about this filter, and gives it its name, is the fact that the low-pass filter is terminated in a 50-ohm dummy load. This arrangement works better than the straight high-pass filter method because it absorbs energy from the rejected band, and reduces (although does not eliminate) the effects of improper filter termination.

Some of the capacitor values are nonstandard, but can be made using standard disk ceramic or mica capacitors using combinations in Table 17.1.

The coils are a bit more difficult to obtain. Although it is possible to use slug-tuned coils obtained from commercial sources (e.g., Toko), or homebrewed, this is not the preferred practice. Adjusting this type of filter without a sweep generator might prove daunting because of interactions between the sections. A better approach is to use toroid-core homebrewed inductors. The toroidal core reduces interaction between the coil's magnetic fields, simplifying construction. Possible alternatives are shown in Table 17.2.

For all coils use wire of a similar UK SWG size to #24 to #30 AWG enamel insulation.

The dummy load used at the output of the low-pass filter (R1 in Figure 17.8) can be made using a 51-ohm carbon or metal film resistor, or two 100-ohm resistors in parallel. In a pinch a 47-ohm resistor could also be used, but is not preferred. In any event, use only

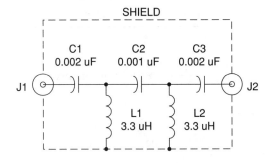

Fig. 17.7 *AM BCB filter.*

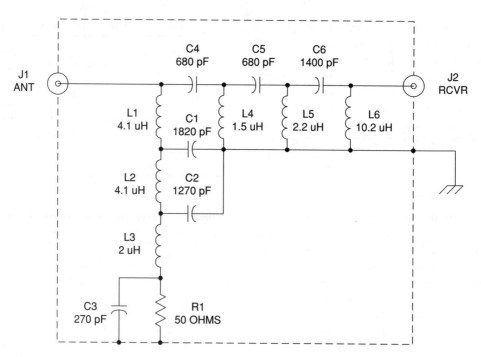

Fig. 17.8 *Absorption type AM BCB filter.*

Table 17.2 Filter Inductance Values

Coil	Value	Core	A_L Value	Turns
L1	4.1 μH	T-50-15 RED/WHITE	135	17
L2	4.1 μH	T-50-15 RED/WHITE	135	17
L3	2 μH	T-50-15 RED/WHITE	135	12
L4	1.5 μH	T-50-2 RED	49	18
		T-50-6 YEL	40	20
L5	2.2 μH	T-50-2 RED	49	21
		T-50-6 YEL	40	24
L6	10.2 μH	T-50-2 RED	49	46
		T-50-6 YEL	40	51

Table 17.3 Filter Reactance Values

Component	X (X_L or X_c)
L1	28.8 Ω
L2	78.4 Ω
L3	38 Ω
L4	28.8 Ω
L5	42 Ω
L6	193 Ω
C1	28.8 Ω
C2	42 Ω
C3	193 Ω
C4	78.4 Ω
C5	78.4 Ω
C6	38 Ω

noninductive resistors such as carbon composition or metal film 1/4- to 2-watt resistors.

If you would like to experiment with absorptive filters at other cutoff frequencies than 3 MHz, then use the reactance values in Table 17.3 to calculate component values.

The exact component values can be found from variations on the standard inductive and capacitive reactance equations:

$$L_{\mu H} = \frac{X_L}{2\,\pi\,F_c} \times 10^6 \text{ microhenrys} \quad (17.1)$$

and

$$C_{pF} = \frac{10^{12}}{2\pi F_c X_c} \text{ picofarads} \quad (17.2)$$

These component values are bound to be nonstandard but can be made either using coil forms (for inductors) or series–parallel combinations of standard-value capacitors.

TRANSMISSION LINE STUBS

At higher frequencies, where wavelengths are short, a special form of trap can be used to eliminate interfering signals. The half-wavelength shorted stub is shown in Figure 17.9. It acts like a series resonant wavetrap and is installed in a similar manner, right at the receiver. The stub is shorted at the free end, so presents a low impedance at its resonant frequency and a high impedance at all other frequencies. The length of the stub is given by:

$$L_{CM} = \frac{1250 \text{ VF}}{F_{\text{MHz}}} \qquad (17.3)$$

where:

L_{CM} is the length in centimeters (cm)
F_{MHz} is the frequency of the interfering signal in megahertz (MHz)
VF is the *velocity factor* of the coaxial cable (typically 0.66 for polyethylene cable and 0.80 for polyfoam; for other types see manufacturer data)

SHIELDING

Shielding is a nonnegotiable requirement of filters used for the EMI reduction task (see Chapter 5). Otherwise, signal will enter the filter at its output and will not be attenuated. Use an aluminum shield box of the sort that has at least 5 to 6 mm of overlap of a tight-fitting flange between upper and lower portions.

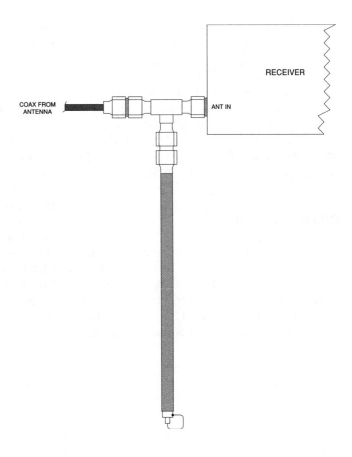

Fig. 17.9
Stub applied to receiver.

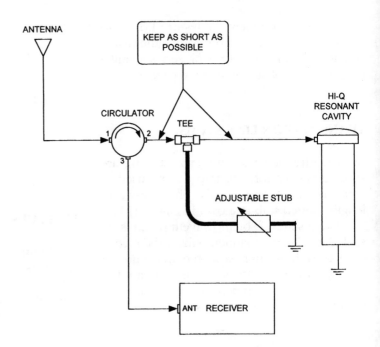

Fig. 17.10
VHF/UHF/microwave receiver
system.

EXPECTED RESULTS

If the correct components are selected, and good layout practice is followed (which means separating input and output ends, as well as shielding the low-pass and high-pass sections separately), then the absorptive filter can offer stop band attenuation of –20 dB at one octave above F_c, –40+ dB at two octaves, and –60 dB at three octaves. For a 3-MHz signal, one octave is 6 MHz, two octaves are 12 MHz, and three octaves are 24 MHz.

DIFFICULT CASES

The UHF-and-up bands are alive with signals, and that makes them prime sources for EMI. In cases where the UHF or microwave receiver is experiencing really serious amounts of EMI, the scheme of Figure 17.10 can be used. In this scheme, a circulator is used to split the

signal into two paths. The first path consists of a stub and a high-Q resonant cavity, which are tuned to the interfering signal. The third port on the circulator is connected to the receiver.

THE SOLUTIONS

So what's the solution? In all of the cases mentioned above, there is one solution that works best of all: prevent the offending signal from entering the receiver. Whether the problem is desensitization from front-end overload, or harmonic generation, or intermodulation problems, the solution is to get rid of the bad signal. In the case of intermodulation products, getting rid of just one of the two signals is all that's needed. A filter in the antenna line ahead of the receiver will work nicely for this purpose.

Chapter 18

Electrostatic Discharge (ESD)

The fundamental aspect of matter that permits the flow of electricity is the existence of free electrons in that matter. The electrical resistance of a material is defined by how many free electrons exist in the material. Metals, for example, have many free electrons, so it is relatively easy to generate a current flow in them. Insulators, on the other hand, have few free electrons, so current flow is proportionally lower. An attribute of insulators is that they do not allow redistribution of electrical charge across their entire surface, so they can permit local electrical charge to build up. This is static electricity and leads to electrostatic discharge (ESD) damage to electronic devices.

Triboelectric generation results whenever sufficient energy has allowed the transfer of free electrons from one insulating material to another by means of contact or friction.

Table 18.1 shows the triboelectric series, which is the electrostatic relationship between the items.

Materials at the top end of the left column of the table have a greater tendency to lose electrons and so are considered "increasingly positive." Figure 18.1 shows what happens when wool is rubbed against hard rubber. Electrons are transferred from the

Table 18.1	Triboelectric Series
Air	Hard rubber
Human hands	Nickel, copper
Asbestos	Brass, silver
Rabbit fur	Gold, platinum
Glass	Sulfur
Mica	Acetate, rayon
Human hair	Polyester
Nylon	Celluloid
Wool	Orlon
Fur	Saran
Lead	Polyurethane
Silk	Polyethylene
Aluminum	Polypropylene
Paper	Polyvinyl chloride (PVC)
Cotton	
Steel	KEL-F
Wood	Teflon
Amber	Silicon
Sealing wax	

wool to the rod. This occurs because the energy level of the valence electrons is raised to the point where they become free electrons and can escape to the hard rubber rod.

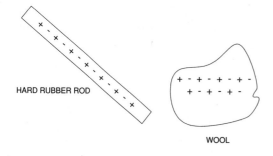

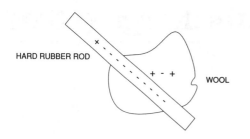

Fig. 18.1   Rubbing hard rubber rod on wool generates static electricity.

Materials can be classified according to the surface resistance in *ohms per square* (ohms/sq). Table 18.2 shows the resistance of materials in ohms per square.

Metals and other conductors are not triboelectric generators because they can redistribute electrical charge received by rubbing over their entire surface, avoiding the local build up of charge. Insulative materials can be charged by triboelectric generation.

The amount of moisture in the air is the humidity, and it can determine the amount of electrical charge that an object can hold. The air resistance can control this charge and will "bleed off" electrostatic charge. This explains why electrostatic discharge will occur more

Table 18.3   Means of Static Generation

Means of Static Generation	Electrostatic Voltages	
	10–20% Humidity	65–90% Humidity
Walking across carpet	35 kV	1.5 kV
Walking over vinyl floor	12 kV	250 V
Worker at bench	6 kV	100 V
Vinyl envelopes for work instructions	7 kV	600 V
Common poly bag	20 kV	1.2 kV
Work chair padded with polyurethane foam	18 kV	1.5 kV

often in the wintertime when the humidity is low and air resistance is high.

The threshold of feeling for electrostatic discharge is around 4,000 volts. Any potential below 4,000 volts cannot be felt, but is still dangerous to electronic equipment. If the static electric spark is heard, then it is in the range of 5,000 to 50,000 volts. It is quite possible for the human body to have an electrostatic charge of 35,000 volts! Table 18.3 shows the charge that can be built up by various activities.

This is compared with various electronic devices' electrostatic voltage tolerance of 30 to 7,000 volts (Table 18.4).

Table 18.4   Static Electricity Susceptibility

Device	Range of ESD Threshold
VMOS	30 to 1,800
MOSFET	100 to 200
GaAsFET	100 to 300
EPROM	100 to 1,000
JFET	140 to 7,000
SAW	150 to 500
Op-amp	190 to 2,500
CMOS (B-series)	
Schottky diodes	
Film resistors	300 to 3,000
Bipolar transistors	
ECL	500
SCR	680 to 1,000
Schottky TTL	1,000 to 2,500

Table 18.2   Resistance of Materials

Type of Material	Ohms per Square
Insulative	Above 10^{14}
Antistatic	10^{9} to 10^{14}
Static dissipative	10^{4} to 10^{9}
Conductive	1 to 10^{4}

ESD EFFECTS

The effects of ESD damage can be devastating to electronic equipment, but it is often misdiagnosed for the following reasons:

1. Failures are analyzed as being due to electrical transients other than electrostatics.

2. Failures are categorized as random, unknown, "manufacturing defects," or other causes.

Very few failure modes and effects laboratories are equipped with the scanning electron microscopes needed to see the damage. Figure 18.2 shows two views of a MOSFET transistor, undamaged (Figure 18.2A) and damaged (Figure 18.2B). Hard faults are immediate and catastrophic failures due to ESD and are immediately apparent. The critical part of the MOSFET device is the thin dielectric insulator between the gate and the substrate/drain/source structure. If enough heat is generated by the static charge, metallic ions will flow into the void created by the ESD, shorting the structure.

Far more devastating are soft faults to electronic equipment. These faults do not appear immediately because the void (Figure 18.2B) is not filled with metal—rather, metal ions migrate into the space from the gate structure over time. Statistics indicate that 90 percent of all ESD may be this type. It is far harder to diagnose because it takes from 90 days to 6 months of operation to fill in the void, with nothing to show for it other than (possibly) reduced performance.

IDENTIFICATION BY CLASS

Electronic components can be classified according to electrostatic discharge capability. Class-1 is the most sensitive, class-2 is less sensitive, and class-3 is the least sensitive. Obviously, class-1 includes MOSFETs and JFETs, or anything on the list that has susceptibility less than 1,000 volts. Class-2 devices

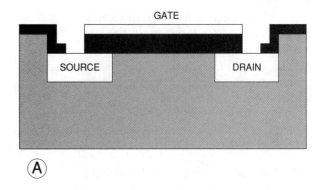

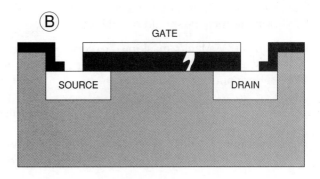

Fig. 18.2
(A) Standard MOSFET; (B) with ESD defect.

have a sensitivity range of 1,000 to 4,000 volts, and class-3 devices have a sensitivity range of 4,000 to 15,000 volts.

ESD CONTROL PROCEDURES

The key to a good program for controlling ESD is found in creating an environment that is protected against generating static electricity. The environment must consider things such as floors, workbenches, materials, equipment, and operating procedures. At the very least, one must equip the technicians and other workers with electrostatic wrist straps, a portable protective work mat, and protection for the parts being placed into the equipment. A manufacturing facility may include an elaborate grounded workbench made of ESD protective materials, humidity controls, conductive flooring and air ionizers, in addition to the above.

An electrostatic location meter will help you determine whether or not static has built up. You can sense voltages from 500 to 50,000 volts with these instruments, depending on model.

In the discussion below the terms "hard ground" and "soft ground" will be used. The term hard ground means that the item is connected to ground through copper or aluminum wiring; the term soft ground means that it is through a resistance.

WORK AREAS

The work area should be controlled-access, which means that people should be kept out unless they are trained. Minimally protect an area at least 1 meter surrounding the work area. The following restrictions should be followed:

> No tapes (3M Magic, masking), except where installed and removed under controlled ionization
>
> No vinyl-covered notebooks or instruction folders
>
> No telephones unless designed for the area

> No plastic note holders, pen holders, calendar holders
>
> No plastic-cased vacuum cleaners or heat guns
>
> No unapproved soldering iron
>
> No unapproved solder suckers

PROTECTIVE FLOORING

Floors are very important in protecting against ESD damage. Only use conductive, static dissipative floors, antistatic carpeting, conductive vinyl, or terrazzo floor tiles. When vinyl tiles are used, the adhesive should be conductive also. Use conductive wax on floors, or leave them unwaxed. Static charge builds up on normal wax.

Painted or sealed concrete floor, or finished wood floors, should be covered with ESD floor mats or treated with a topical antistatic compound, if no floor mats are available. In any event, the floor should be tested periodically to determine if it is still conductive.

Use of a grounded floor is not terribly useful if the individual worker is not soft grounded also. Use conductive shoe covers, conductive shoes, heel grounders, or some similar device to ground the body of the worker. Conductive work stools or chairs should also be used.

WORKBENCHES

Figure 18.3 shows a workbench set up for ESD protection. It is set on conductive flooring or a conductive mat and is made of conductive materials.

There is a protective wrist strap on the bench that grounds the worker. It should have a resistance that limits current flow to 5 mA, the threshold of perception, at the voltages that will be worked on the bench. For 240 volts, that means the resistance should be higher than 240/0.005 = 48,000 ohms.

Two types of ESD wrist strap are used: carbon-impregnated plastic and insulated metal conductor. The carbon-impregnated type

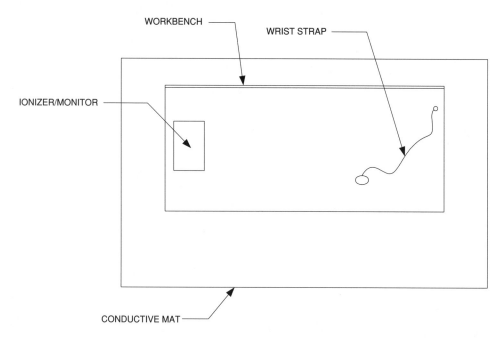

Fig. 18.3   Workbench.

distributes the resistance over its entire length, so it should be insulated to guard against touching hard grounds. The insulated metal conductor type contains a built-in resistor. The strap should include an alligator clip, snap, or other form of quick disconnect to permit the user to disconnect in emergencies or when leaving the work area.

There is an ionizer and monitor on the bench in cases where manufacturing or very sensitive work is taking place.

EQUIPMENT

The equipment used on the ESD workbench should be designed for the purpose. Soldering equipment should be three-wire types with grounded tips. The resistance of the tip to hard ground should be 20 ohms or less so that the voltage buildup will be less than 15 volts. Solder suckers used should be ESD types, which means at least they will have conductive tips.

All exposed metal surfaces on test equipment should be hard grounded. However, the test equipment should not be set directly on the workbench's conductive surfaces for fear

of rendering the surface, and the person using the equipment, at hard ground potential. Ground-fault interrupters should be used on all outlets on the workbench.

CLOTHING

People handling ESD-sensitive electronics products should be properly attired to do so. They should wear long-sleeved ESD smocks or close-fitting short-sleeve shirts. The use of ESD protective aprons is encouraged. Never use smocks or gloves made of common, ordinary plastic, as it is not ESD protective.

ESD-PROTECTIVE MATERIALS

Protective materials should be used to protect electronic products containing ESD-sensitive components. Chips should be mounted on those foam carriers and not removed until they are ready for use. The finished printed circuit cards should be stored in ESD bags, wrappings, or boxes.

Chapter 19

Regulatory Issues

In this chapter we are going to look at some of the regulatory issues regarding EMI/EMC. Governments have long considered EMI/EMC to be among their functions, and as such have issued a number of standards and regulations.

The military has long recognized a severe EMI/EMC problem. Their equipment must perform, and perform well, in the presence of fields as strong as 10,000 volts per meter (V/m). The military has a couple of specifications on EMI/EMC. MIL-STD-461 and MIL-STD-462 (in various versions) have traditionally been used for the EMI/EMC requirements, but have recently been replaced by MIL-STD-461E (which combines '461 and '462).

The European Community issued directive 89/336/EEC to cover EMI/EMC. That directive carries criminal penalties and requires all electronic products to be tested for compliance. The European Community EMI/EMC directive requires every manufacturer or importer of electronic apparatus make a "Declaration of Conformity" for the unit. The "CE" mark must appear on the equipment at least 5 mm high and be indelibly affixed to the unit (ever wondered what that "CE" mark meant?). Failure to conform to the directive can lead to legal sanctions.

The International Electrotechnical Commission and the Organization for International Standards (issuers of the famous "ISO-xxxx" series of documents) work closely together on EMI/EMC issues. Both are located in the Swiss city of Geneva. The International Special Committee of Radio Interference (CISPR after the French) is the IEC committee that works mostly on EMI/EMC issues. The European Telecommunications Standards Institute (ETSI) and the European Organization for Electrotechnical Standardization (CENELEC) are two organizations that cooperate with the IEC (even though their mandate is solely European) on EMI/EMC matters. In the United States, the American National Standards Institute (ANSI) and the Institute of Electrical and Electronics Engineers (IEEE) are the responsible parties. The Federal Communications Commission regulates radio reception in the United States and as such is concerned with EMI/EMC of telecommunications equipment.

The technical vocabulary of EMI/EMC is given in a document called IEC-50. It has definitions given in three languages (English, French, and Russian). It also has a number of terms in Dutch, German, Italian, Polish, Spanish, and Swedish.

IEC-801 defines compliance of electronic units. IEC-801-2 defines testing for electrostatic discharge (ESD), while IEC-801-3 defines equipment for industrial process management and control. Another important IEC document is IEC-1000.

There are a number of standards in EMI/EMC besides 461/462. If a standard begins with a "BS" it is a British standard, while an "EN" prefix denotes a European standard ("norme" is the French word for "standard" and is the "N" in "EN"). Some standards are issued as both BS and EN types. For example, BS-800 (on radio interference) is also EN-55014. The "14" in EN-55014 denotes that the standard is derived from CISPR 14.

EN-50081 (Emissions) and EN-50082 (Immunity) are generic standards for equipment that is not covered by other standards. If other standards exist, then they take precedence over EN-50081 and EN-50082.

BS-6527, also known as EN-55022 and CISPR 22, specifies that between 30 and 230 MHz the electrical field 10 meters from digital equipment must not be 30 μV/m. That is a lot of interfering signal to be radiated into a receiver! It might not be sufficient for on-channel interference, yet it is the most severe civilian standard in the world.

The FCC publishes radio interference standards in the United States, with input from ANSI and IEEE. Part-15 CFR 47 (Code of Federal Regulations) defines the EMI/EMC world of unintentional interference. It recognizes two main classes: Class-A for business or industrial use, and Class-B for domestic use. Class-B is stricter than Class-A because radio receivers and television receivers are susceptible to EMI/EMC. Publication FCC/OET MP-4 describes how to measure the interference emitted by computers and other devices.

Europeans use the same "class-A/B" designation as the United States. In Europe, a Class-B device is (generally speaking) one that operates from a 13-ampere outlet.

There are three possibilities for obtaining the CE mark. The first is conformance of the equipment to a standard. In the absence of a standard covering the equipment being sold, the importer or manufacturer must create a technical construction file. In Great Britain, the file must be approved by a NAMAS-approved accredited laboratory. In Germany there is a group called TÜV (for Technischer Überwachungs Verein) that is the "competent body" to do the examination of equipment.

In the case of radio transmitters there is a method for complying with the law, and that is approval by the British Approvals Boards for Telecommunications (BABT). The BABT is also the approval unit for telecommunications terminal equipment connected to telephone lines.

ANECHOIC CHAMBERS AND OATS

Conformance testing is done in anechoic chambers (Figure 19.1) or on open area test sites (Figures 19.2 and 19.3 show a typical OATS that conforms to EN-55022). If the funds and facilities are available, then an indoor anechoic range (Figure 19.1) is preferred. The cones shown in Figure 19.1 are 2.5 meters in length at 30 MHz, although they are smaller for VHF/UHF ranges.

The data collected are far more useful than the data collected out of doors, especially off-the-air. The indoor anechoic range is simply a more easily controlled, predictable environment, despite being more expensive.

The indoor range consists of a room lined with materials that absorb radio waves. Usually in the form of pyramid-shaped wedges, various sizes of absorbers are placed over all surfaces of the chamber. The room is inside a *Faraday cage*. This shielding provides protection against both signals from outside interfering with the test, and reflections of the reference signal from structures outside of the chamber. A point source, such as a small loop at UHF and below, or a horn radiator at microwave frequencies, is positioned at one end of the room.

The same type of range can be used for antenna patterns. The antenna is mounted on a rotatable pedestal. Both elevation and azimuth plots can be made by the simple expe-

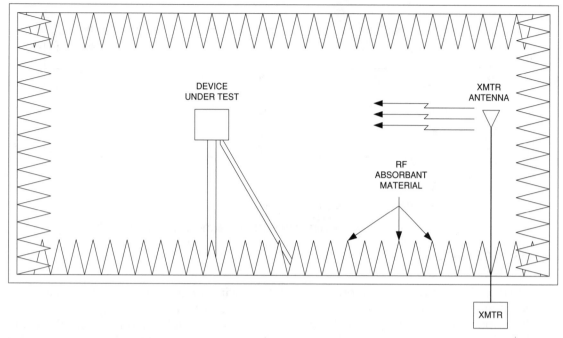

Fig. 19.1 *Anechoic chamber.*

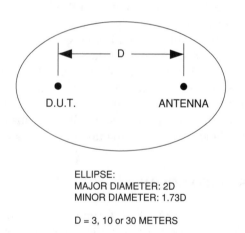

ELLIPSE:
MAJOR DIAMETER: 2D
MINOR DIAMETER: 1.73D

D = 3, 10 or 30 METERS

Fig. 19.2 *OATS Form 1.*

dient of mounting the antenna in the correct plane for each test.

In order to prevent secondary reflections from the pedestal, blocks of absorbent material are often used to block the view of the pedestal from the radiator.

In antenna measurements, the polar pattern of the antenna is created by using a servomotor or position transducer to rotate a chart recorder that measures the output strength of the receiver. The result is the familiar antenna azimuthal or elevation patterns.

In antenna measurements, modern versions of the indoor range use a computer rather than the polar plotter. The output of the receiver is a voltage indicating signal level, so it can be A/D converted for input to the computer. The position transducer can also provide data to the computer. Once these data are in a plotting program, it can draw the polar plot.

One advantage of the computer approach is that a static profile of the chamber can be made by mounting a reference antenna in place of the antenna under test (AUT) or device under test (DUT). The plot can then be made and note taken of any reflections or other anomalies that exist in the chamber. These data can then be compared with the data for the AUT or DUT at each angle, with the AUT data adjusted to account for directional differences.

An advantage of the indoor test facility is that it allows test engineers and technicians

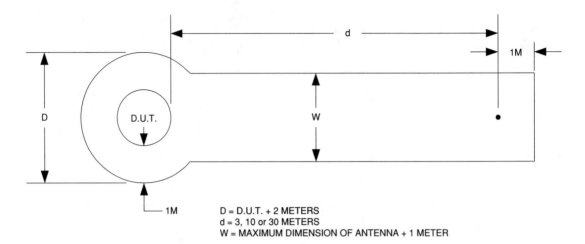

Fig. 19.3 *OATS Form 2.*

to rotate the DUT in any attitude required for the measurements.

The open area test site (OATS) is used out doors. An EN-55022 compliant OATS is shown in Figures 19.2 and 19.3. The typical OATS is an ellipse with major and minor axes.

The device under test (DUT in Figure 19.2) and antenna for the test signals are on the metal ellipse. The dimension D may be 3, 10, or 30 meters depending on the size of the DUT. The major axis is $2D$ and the minor axis of the ellipse is $1.73D$.

The nature of the OATS ground screen depends on the frequencies for the ellipse. It can be a screen, but the "holes" in the screen should not be a major part of one wavelength at the lowest frequency of operation. The holes in the mesh should be 33 mm, which is one-tenth wavelength at 1 GHz. That is usually not a problem, but the minimum and maximum frequencies of the OATS must be considered.

For straight EMI/EMC testing, the OATS is sufficient because reflections don't bother EMI/EMC tests as much as they do antenna patterns. Nonetheless, care must be taken to locate the OATS where there are no over-head wires or other structures that reflect radio-frequency energy used in the test.

The test site should have an RF-friendly structure for the electronics and personnel involved. A glass-reinforced polyester (GRP) facility is often used to avoid RF reflections. A better approach is to put the control electronics and control room below ground.

A second form of OATS is shown in Figure 19.3. It places the DUT in a circular area that is at least one meter on any dimension from the DUT. The distance d is 3, 10, or 30 meters; D is DUT plus 2 meters; W is the maximum dimension of the antenna plus 1 meter.

SCREENED ROOMS

A decent job of EMI/EMC testing can be done inside a screened room, rather than an anechoic chamber. The cost of the anechoic chamber is prohibitive for many users (although one can be rented for a few thousand dollars per day). The screened room is basically a Faraday cage. The size of the screened room is $4 \times 5 \times 3$ meters. In IEC-801-3, however, the size of the screened room will be increased to $6 \times 5 \times 3$ meters.

Automotive Interference Solutions

ARRL TECHNICAL INFORMATION SERVICE INFORMATION ON AUTOMOBILE–RADIO INTERFERENCE (BOTH WAYS)

File: rfiauto.txt

Updated: November 17, 1995

Author: Michael Tracy, KC1SX (e-mail: tis@arrl.org)

Reprinted from: September 1994 *QST* "Automotive Interference Problems: What the Manufacturers Say" and March 1995 QST "Lab Notes: Mobile Installations and Electromagnetic Compatibility"

Copyright 1995 by the American Radio Relay League, Inc.

All rights reserved.

For your convenience, you may reproduce this information, electronically or on paper, and distribute it to anyone who needs it, provided that you reproduce it in its entirety and do so free of charge. Please note that you must reproduce the information as it appears in the original, including the League's copyright notice.

If you have any questions concerning the reproduction or distribution of this material, please contact Michael Tracy, American Radio Relay League, 225 Main St., Newington, CT 06111 (e-mail: tis@arrl.org).

NOTE!
This information package provides introductory information only. Additional information on this subject and related topics can be found in the following ARRL publications:

The ARRL Handbook (#1735)

Radio Frequency Interference: How to Find It and Fix It (#3754)

The ARRL RFI Book: Practical Cures for Radio Frequency Interference

AUTOMOTIVE INTERFERENCE PROBLEMS: WHAT THE MANUFACTURERS SAY

Installing mobile radios in your new megabucks car can be a frightening proposition—especially if RF from your rig could damage your shiny new roadster, voiding the warranty! Want help? Read on!

By Ed Hare, W1RFI, ARRL Laboratory Supervisor

In the good old days, things were simple. If you wanted to install a mobile radio in your car, your primary considerations were mechanical: where to place the antenna, where to mount the radio and how to route the wires so they didn't interfere with the family use of the car.

There were some incompatibility problems to solve, such as electrical RF noise from the ignition system, but the proper application of cures (resistive wires and plugs, or, in extreme cases, a filter for the distributor and its wiring) almost always resulted in a successful installation.

As automotive designs evolved, however, installing mobile radio systems became more complicated. Remember how computers used to be house-sized monoliths? In the early 1980s, microprocessors and their associated circuitry became small enough (and inexpensive enough) so car manufacturers (and many others) could use them to control many different functions.

In the early 1980s, electronic control modules (ECMs) became standard in most cars. There were sporadic reports of interference problems to and from these devices, but most hams were able to find a work-around. Soon it became possible to use microprocessors to accomplish additional automotive functions, ranging from engine control to anti-lock braking to airbag deployment. Some manufacturers even use "slave" microprocessors to control things such as rear-end lights, running only one cable to the back instead of an entire wiring harness, using the "slave" microprocessor to execute proper tail-light sequence (brakes, signaling, etc.).

Sure enough, these electronic marvels came with a price!

The more complex things become, the more likely it is that things will go wrong.

Every microprocessor has a clock oscillator, and the circuitry uses digital signals for processing and control. These digital signals are square waves. They're perfect for digital circuits, but rich in harmonics.

FCC regulations (Part 15) specify the amount of interference that can be generated by these "unintentional radiators." The regulations are adequate to protect other radio services, such as TV reception in nearby homes, but they're not intended to protect against interference to radio receivers installed in the vehicles, broadcast band or otherwise.

In addition, the vehicle electronics can also be affected by strong electromagnetic fields (EMFs). These fields can be caused by nearby transmitters, transmitters installed in the vehicle, high-voltage power lines, and so on.

Most manufacturers created electromagnetic compatibility (EMC) departments to deal with testing and design issues, assuring compliance with federal regulations and compatibility with factory-installed equipment.

In addition, the industry has voluntarily developed standards that apply to many EMC technical issues (see the sidebar on the Society of Automotive Engineers). There is even a standard that applies to installed transceiver equipment. (The ARRL is a voting member of the SAE Committee working on these standards.)

So far, things look pretty good! New technology has made cars less expensive, more reliable and less polluting. Federal regulations control the amount of interference that cars can generate, automobile manufacturers have created departments to solve the problems and the entire industry has formed committees and developed standards to help make things right.

As automotive designs evolved, ham gear did, too. More and more ham transceivers were capable of operating from a 12- to 14-volt supply, so mobile operation became more popular than ever.

Unfortunately, this rosy picture was spoiled by an unexpected phenomenon; radio transmitting equipment, sensitive receivers and automotive electronics didn't always work well together.

As if this weren't bad enough, EMC problems usually take place on a two-way street (all puns intended). Just as vehicle electronics can interfere with installed radio equipment, even low-power transmitters can interfere with vehicle electronics.

Details are not forthcoming, but urban legends abound about vehicles that would stall or lock their brakes near high-power transmitters, or about hams who could stall nearby vehicles on the highway by keying up high-power transmitters. (The ARRL staff has amassed a fair collection of anecdotal reports, none of which describe this problem firsthand.)

The legends may or may not be true, but vehicle manufacturers know that fields of up to 300 V/meter can be found on our highways and byways; automobile electronics must continue to function when drivers whiz past Voice of America transmitter sites!

Car companies have worked to ensure that their vehicles do not interfere with factory-installed equipment and do not keel over near VOA-class transmitters, but it's clear that manufacturers do not always pay attention to compatibility with after market equipment, including transmitters and receivers for various radio services.

When hams installed transceivers in their cars, things didn't always work as planned. As the number of automotive microprocessors grew, the potential for the umpteenth harmonic of the clock oscillator falling on a favorite repeater channel also grew. Add to this the possibility of noise from sophisticated ignition systems, motor noise from wiper-blade, electric cooling fan or fuel-pump motors, and even the vehicle's factory-installed broadcast-band radio receivers and you have a potential for electromagnetic incompatibility.

What's worse, some vehicle electronics are susceptible to RF fields generated by mobile transceivers. This susceptibility ranges from the minor annoyance of having a dash light come on in step with the transmitter, to the major annoyance of having the vehicle's microprocessor lose its mind, resulting in a dead car.

The ARRL has not received any reports about interference to safety devices such as anti-lock brakes or air bags, but this type of interference is still possible, especially if good installation techniques are not used.

WHEELS START TO TURN

In early 1992, John Harman, W8JBH, wrote an item for *QST*'s Correspondence column pointing out the problems he was experiencing in trying to get Toyota to help him with an interference problem he was having with his 1992 Camry.[1]

In Harman's case, his 2-meter transceiver had resulted in temporary damage to the car's ECM. John, as of that date, had been unable to get any concrete information from Toyota about the proper installation of transmitting equipment in his vehicle.

Since then, we've heard from other hams who experienced similar interference. The May 1992 issue included a few more tales of woe. In addition, the ARRL Technical Information Service and RFI Desk heard from a few dozen folks who had been having some sort of EMC problems with their cars, dealers or manufacturers.

This is, of course, just the tip of the iceberg; most hams simply do not report their interference problems: not to the FCC, the involved manufacturers, or the ARRL.

The situation was confusing. After hearing tales of blown ECMs and voided warranties, hams were afraid to install mobile transceivers in their cars. As interference reports increased, the Lab decided to look into the matter.

ARRL SURVEYS AUTO MANUFACTURERS

As ARRL Senior Lab Engineer and all-round EMC guru, the task fell into my lap. "No problem!" I said. "We can ask car manufacturers

to tell us all about their cars and policies."
I drafted a letter asking the following
questions:

- "How does your company resolve electromagnetic compatibility (EMC) problems that result from installed (or
  nearby) transmitter operation, or when
  vehicle electronics cause interference
  to installed (or nearby) radio receivers?

- "Have you published any service bulletins that relate to radio transmission
  or reception, or electromagnetic interference (EMI)?

- "If customers have problems with EMC
  or EMI that cannot be resolved by the
  dealer, who should the dealer or customer contact for additional assistance?
  Have these contact people been specifically trained in EMC and EMI mitigation?

- "May we make the information you
  provide available to our members?"

We also outlined some typical Amateur
Radio installations, citing power levels of
100W on HF, 50W on VHF and 10W on UHF.

I sat back and waited for a flood of responses from a field of eager manufacturers. I
knew that each would tell me that it was okay
to put transmitters in their cars and that if hams
had problems, dealers and customer service
people would be glad to help them out.

The replies would lead to an article
summarizing the responses that would clearly explain how to install a radio in each type
of car and reap the praise from all of our
grateful members. NOT! (Well, not exactly,
anyway.)

After 60 days, we'd heard from only a
handful of manufacturers. The first round of
responses didn't look very useful, with the
letters ranging from "We have never had a
problem with radio installations," to "It's not
our fault—ever!" The latter is paraphrased,
but not too far off the mark.

Things weren't going to be as smooth as
I had hoped. I waited impatiently another
month and sent off a follow-up letter, asking
what happened to the first letter, and pointing

out how the manufacturer was going to look
if it did not respond and were listed in a national magazine as having ignored two letters.

Over the next 90 days, the answers
trickled in. In many of the letters I could
clearly see the mark of the manufacturer's
marketing, public relations and legal departments. The caveats were rampant, and in
most of the letters the disclaimer that if any
after market equipment, including radio
transmitters, caused any damage to the vehicle it would not be covered under warranty.

One company said the answers to the
questions I had asked were proprietary!
Even worse, some companies didn't respond at all! In most cases, I finally got a
response by calling the respective public
relations departments.

To partially offset the number of companies that didn't respond, one company
sent two answers—completely contradictory,
of course.

The mix-up was ultimately resolved,
but it demonstrates that a big part of the EMC
problem may involve poor communication.

A company can have an excellent EMC
facility, program and policy, but if its information isn't widely distributed to dealers and
regional offices, it does little good.

If *QST* received contradictory responses
from the same building within the same company, imagine how difficult it could be to get
accurate information from corporate and engineering policy makers at the factory to dealers
who have to actually solve the problems.

So far, the overall manufacturers' response hasn't been that good. It isn't all bad,
however. I noticed several call signs in the
signatures of the letters I received. Not sur-
prisingly, these were among the more useful
answers.

A few of the companies came quite
close to my ideal, giving us solid information
and telling us that their dealer-support network or factory specialists will help hams
with compatibility problems!

Toyota, the company whose customer
really started the ball rolling, did respond after
much prompting (not unlike John Harman's
experience). And after all was said and done,

Toyota's response was actually positive. It took a bit of time, but for the 1994 models, the 10-W power output limitation mentioned in the 1992 Service Manual has been upped to 100 W. According to Toyota's Customer Service Department, a few of the 1994 manuals were not updated, but the current limitation of 100 W applies across the board.

General Motors has long been proactive in the EMC field, maintaining a complete EMC facility (like many other automobile manufacturers), publishing a complete and useful installation guide and maintaining a high profile in the professional and amateur EMC press.

Even more encouraging is the fact that several companies told me privately that my letter and article have prompted them to develop their own EMC guidelines and to clarify company policies about EMC.

Unfortunately, there have been problems even with some of the companies that have clear policies and installation guidelines. In several cases, hams who have reported problems tell us that customer-service people or dealers made decisions that were different from the information that we received in writing from the manufacturer.

When I called the Customer Assistance Center at Pontiac, as a follow-up to a member who called me, at first I was told the same thing as the member—that they had no information about how to install radios in cars. Everyone I spoke with at the assistance center insisted that the dealer is the only source of information.

Remember, GM is one of the good guys! They have an active EMC department, published installation guidelines and clear policies. Unfortunately, the process often breaks down.

All in all, the problems are not surprising. Many auto manufacturers are giants, and it must be difficult to maintain a clear policy in all areas across such a diverse and spread-out business structure. On the other hand, our original letters were sent directly to each company's customer-service contact address.

Of course, all of the policies could be swept away by the new model year! As each new model is designed and built, its EMI susceptibility presumably will be different than its predecessor. Because most manufacturers won't guarantee that a properly installed transceiver will not interfere with a vehicle's proper performance, we could never keep track of the changing policies and applications.

Clearly, manufacturers have a way to go before the policies and standards formulated by their hard-working EMC engineers and EMC committees can make a difference for the average ham. The ARRL will continue to work toward that goal—and you can help! If you've had interference problems with your automobiles, or good or bad experiences with your dealer in getting those problems resolved, report this to the manufacturer at the address we list near the end of the article. This will help manufacturers to stay on top of the types of problems their customers are experiencing.

Send a copy of your letter to ARRL Headquarters, RFI Desk, 225 Main St. Newington, CT 06111-1494. I will collect the letters, combine them with what we already have, and send them en masse to the key players at automobile manufacturers and their standards committees.

Let's see how we can make a difference. If all else fails, contact me at HQ with your automotive EMC problems.

WHAT THE MANUFACTURERS SAY

Now, as promised, a description of each manufacturer's policy statement. The addresses and telephone numbers are the manufacturers' suggested contacts or are the addresses on the letterheads of the letters we received from manufacturers. All of this information is based on the statements made by the vehicle manufacturers. Manufacturers not listed were not contacted. If the manufacturer answered our question about who to contact to resolve problems, we have included that in their response. If not, contact your dealer or the manufacturer's customer-assistance network. Most manufacturers suggest that customers contact their dealers, who have access to the normal problem-resolution workers through zone or regional offices, which, in turn, contact the factory for support with difficult problems.

BMW of North America Inc.
Customer Relations
300 Chestnut Ridge Rd.
Woodcliff Lake, NJ 07675
201-307-4000

BMW does not test their cars for the installation of the types of radios we mentioned (ham HF, VHF and UHF transmitters). They cannot comment on the transmitters' compatibility with BMW products. However, the electronic systems in their vehicles are designed to be protected from EMI sources outside the vehicle. A specific repair would not be covered under warranty if it were determined that damage was caused by the installation or operation of any non-approved after market accessory.

Chrysler Corporation
Customer Satisfaction and Vehicle Quality
26001 Lawrence Ave.
Centerline, MI 48015

Chrysler has an extensive EMC program, involving design, testing and active participation in the development of national and international standards on EMC for the automotive industry. They support the power levels described in SAE J551/12 (typically 100 W—see the sidebar) for 1992 through 1995 model years. On new vehicle orders, specify the JLW sales code to obtain a vehicle with a suppressed ECM.

The owner's manual of a new Chrysler product contains a summary of their EMC policies. The design of the vehicle provides immunity to radio-frequency signals. Two-way radio equipment must be installed properly by trained personnel. Power connections must be made directly to the battery, fused as close to the battery as possible. Antennas should be mounted on the roof or rear of the vehicle. The antenna feed line must be shielded coaxial cable, as

short as practical and installed away from vehicle wiring. Ensure that the antenna system is in good order and is properly matched to the feed line.

Chrysler has published a comprehensive installation guideline, expanding on the guidelines in the owner's manual. Any Chrysler dealer can supply or order a copy (order no. TSB-08-31-94). The guideline describes in detail the mechanical, electrical and RF requirements that need to be considered for a successful radio installation.

All of this withstanding, Chrysler warranty does not cover any non-Chrysler parts or the costs of any repairs or adjustments that might be needed because of the use or installation of non-Chrysler parts. The owner's manual details the policy that should be followed to resolve any customer problems that can't be corrected by the dealer. Chrysler had a booth at the 1994 Dayton HamVention, answering customer questions, passing out literature, and soliciting customer feedback.

Ferrari North America, Inc.
Corporate Office
Director of Parts and Services
250 Sylvan Ave.
Englewood Cliffs, NJ 07632
201-816-2650

Ferrari North America is a distributor for its parent company in Italy. It did not have much information available, but stated that any vehicle damage caused by the installation of any after market equipment would not be covered under warranty. At press time, the New Jersey office was still awaiting information from engineers in Italy.

Fiat Auto R&D USA
39300 Country Club Dr.
Farmington Hills, MI 48331-3473
810-488-5800

Fiat Auto R&D USA is also a distributor for its parent company in Italy. They did not have any information about the installation of transmitters in their vehicles or how to resolve any EMC problems. They suggested that we contact one of their engineers at their factory in Italy.

Ford Motor Company
Ford Parts and Service Division
Public Affairs
300 Schaefer Rd.
PO Box 1902
Dearborn, MI 48121
313-446-8321

The electronic modules and entire vehicle are subjected to conducted/radiated immunity levels testing. These tests are designed to reflect the use of 100-W class transmitters installed in the vehicle and verify that their operation causes no damage to any system and that critical safety systems will not be functionally affected during exposure.

Tests are also done to minimize interference with radio reception.

However, Ford cannot guarantee freedom from interference from all possible installation configurations. The installation of such equipment does not necessarily void the warranty, but if the installation is determined to have caused damage, such damage would not be covered. (Ford engineers say they do test for the installation of on-board transmitters.)

Ford is not aware of any damage problems from Amateur Radio equipment. If problems are experienced, first contact the installer or supplier of the ham equipment to see if a different installation procedure will correct the situation. If that doesn't work, contact your dealer, who has access to Technical Service Bulletins (TSB) and technical support. There is a TSB (93-15-6) on the installation of a filter to reduce interference from the electric fuel pump (reports received at ARRL are mixed; for some hams it worked, for others it didn't). According to Ford, these TSBs are only recommendations and are not guaranteed to work.

In 1987, the Ford Electrical and Electronics Division wrote an installation guide (Tech Letter EED-6-031 -N). This was not mentioned in Ford's 1992 letter to us, so it is probably obsolete.

General Motors

Each GM subsidiary has its own customer-service network and TSB numbers for installation guidelines. These guidelines represent the only official EMC policy of General Motors. The following is taken from GM installation guidelines:

"Certain radiotelephones or land mobile radios or the way in which they are installed may adversely affect vehicle operations such as the performance of the engine and driver information, entertainment and electrical charging systems. Expenses incurred to protect the vehicle systems from any adverse effect of any such installation are not the responsibility of the General Motors Corporation." The following are general guidelines for installing a radiotelephone or land mobile radio in General Motors vehicles. These guidelines are intended to supplement, but not to be used in place of, detailed instructions for such installations which are the sole responsibility of the manufacturer of the involved radiotelephone or land mobile radio.

"If any vehicle-radio interaction exists after following these guidelines, check current service bulletins for resolution of the customer problem. If there is no bulletin that covers the customer problem, call your technical assistance group." [This is apparently an instruction to the dealer.—Ed.]

Locate the transceiver for remote radios on the driver's side of the trunk. One-piece transceivers should be mounted under the dash or on the transmission hump. Don't mount any accessory in the deployment path of the air bag. Mount antennas in the center of the roof or center of the rear deck lid. Use a high-quality coaxial cable, located away from vehicle electronics. Tune the antenna for an SWR < 2:1. Obtain power directly from the battery itself, properly fused, for both positive and negative leads. Use connector kit 1846855 (GM and AC-DELCO) to connect to the battery terminals. Route all wires through grommeted holes in the front bulkhead.

As an editorial aside, General Motors and its various subsidiaries have long been active in the EMC field. They have had a booth at the annual Dayton HamVention to gather customer feedback and to issue information about their automobiles. They were one of the first companies to publish installation guidelines.

In general, hams should not experience problems with the installation of transmitters in General Motors vehicles. Their guidelines outline the approved methods to install transmitters in their cars. GM design engineers test the installation of on-board transmitters.

Honda, American Honda Motor Co., Inc
National Consumer Affairs
1919 Torrance Blvd.
Torrance, CA 90501-2746
310-783-3260

According to Honda, the installation of amateur transceivers has not presented any problems. If a customer experiences a problem, Honda will refer the customer to the original installer or manufacturer of the installed equipment. The installation of an after market transmitter would not necessarily void the warranty, but if the installation were not done properly or damage were caused by a defective after market component, it would not be covered.

Hyundai Motor America
Consumer Affairs
10550 Talbert Ave.
Fountain Valley, CA 92728-0850
800-633-5151

Hyundai has an extensive EMC test facility at its factory in Korea. They have not experienced many problems with properly installed, medium-power transmitters. Their customer-service engineers have been able to work successfully with their dealers to straighten out any problems that have been encountered. They do not have any published installation guidelines, but suggest that a proper installation would keep power and

antenna cables well away from the ECM and its wiring, obtain power directly from the vehicle battery and ensure that the antenna were properly tuned and installed, well away from the ECM. The ECM is located just behind the kick panels on the left or right side of the vehicle. Customers who experience problems should first carefully evaluate the installation, then, if necessary, contact their dealer. Dealers have access to Regional Offices and the National Technical Services Department for assistance in resolving customer problems. This information was supplied in 1992. As of press time, they had not yet answered a request to verify the information as being applicable to the current model year.

Isuzu, American Isuzu Motors, Inc.
Customer Relations Department
13181 Crossroads Parkway N
City of Industry, CA 91746-0480
800-255-6727

The installation of radio transmitters has not been tested by the parent company in Japan, so Isuzu does not have any information. Their policy letter is clear: "Any modification to your vehicle could affect its performance, safety or durability; void the warranty; or even violate government regulations." [Apparently this policy was intended to cover more than just ham radio transmitters.—Ed .]

In a conversation with ARRL, the Customer Relations Manager was quite clear; they deal with only unmodified Isuzu vehicles. Problems with the installation of transmitting equipment fall outside the scope of their customer service responsibilities.

Mazda Motors of America, Inc.
Customer Relations Manager
PO Box 19374
Irvine, CA 92718
800-222-5500

No answer was received to two letters and three faxes sent to Mazda's "Customer Relations Manager." A letter to the Public Relations Office prompted a phone call from Mazda's Customer Service Coordinator, who promised an answer within "two weeks." As of press time (about two months later), the answer had not been received.

Mercedes-Benz of North America
Customer Assistance Center
One Mercedes Dr.
PO Box 350
Montvale, NJ 07645-0350
800-FOR-MERCEDES

Mercedes-Benz says that it is generally acceptable to install moderate-power transmitters in their cars, provided the installation guidelines outlined in their Service Information MBNA 54/35 are followed. Customers having problems should first contact their dealer, who will then contact the nearest Regional Office. The Regional Offices have received training and information about how to resolve interference problems. Dealers or customers can also contact the Customer Assistance Center at the toll-free number listed above. Mercedes-Benz is committed to providing reasonable technical assistance to their customers. They have an extensive laboratory and field EMC testing program.

Mitsubishi Motor Sales of America, Inc.
6400 Katella Ave.
Cypress, CA 90630-5208
714-372-6000

The installation of after market equipment does not necessarily void the warranty, but if the after market equipment causes any damage to the vehicle, that damage will not be covered under warranty.

Nissan Motor Corporation in the USA
Consumer Affairs
PO Box 191
Gardena, CA 90247-7638

All requests about EMC problems should be directed to the above office. The staff is not specifically trained in EMC, but will direct inquiries to the correct departments.

Poor installations that do not follow the Nissan installation guidelines and high transmitter power can cause malfunctions and/or damage to electronic control systems of any vehicle. While the installation of a radio in and of itself will not void the warranty, any expenses incurred in protecting the vehicle's electronics from a radio installation are not the responsibility of Nissan Motor Corporation.

They did supply a copy of an installation guideline. The guideline emphasized the following points:

No service bulletins have been published.

The radio equipment must be type accepted by the FCC. [Most amateur equipment doesn't require type acceptance, so this doesn't usually apply to hams.—Ed.]

The radio-equipment power connections should be run directly to the vehicle battery. The antenna and power connections should not be routed along with any vehicle wiring, and should cross vehicle wiring at right angles. If possible, route the antenna and power cables in contact with the vehicle body. Use quality coaxial cable (95% or better shielding) with proper antenna connections. The SWR must be below 1.5:1. The antenna should be connected directly to the vehicle's body, as far away as possible from all on-board vehicle electronics modules.

Moderate power levels may be used in Nissan vehicles. Moderate power is defined as: <100 W below 500 MHz; 10 to 40 W between 500 and 1000 MHz; 1 to 10 W above 1000 MHz. In addition, they have a figure depicting the recommended antenna locations.

Peugeot Motors of America
One Peugeot Plaza
PO Box 607
Lyndhurst, NJ 07071
201-935-8400

Peugeot has ceased production of US models after the 1992 model year. They recommend against the installation of transmitters in their vehicles. Their ECMs are not shielded adequately to protect against the resultant amounts of EMI which may interrupt or damage their operation. If damage to a Peugeot vehicle is determined to be caused by an outside influence, the damaged component would not be covered under warranty.

Porsche Cars North America, Inc.
100 West Liberty St.
PO Box 30911
Reno, NV 89520-3911
702-348-3000

Porsche "prohibits the installation of radio transceivers and any after-market electric equipment in new Porsche vehicles." They "discourage the addition or use of non-Porsche supplied telephone equipment in the vehicle." If transmitters are operated within their cars, control or warning lamps may light up with no apparent reason, or malfunctions may occur in other electronic components.

Saab Cars, USA
Consumer Assistance Center
PO Box 9000
Norcross, GA 30091
800-955-9007

All Saab vehicle designs have been EMC tested. Individual questions about frequency and power can be answered by their Customer Assistance Center. All inquiries and problems are handled on an individual basis. They have no record of any EMC problems with Saab cars. The installation of after market accessories will not void the vehicle warranty, but if the installed equipment causes any damage, that damage will not be covered under warranty.

Subaru of America
Owner Service Department
PO Box 6000
Cherry Hill, NJ 08034-6000
609-488-3278

According to Subaru, they have not experienced any problems with the installation of on-board transmitters. If any problems are experienced, the customer should contact the dealer. If the installation of any after market component causes damage to the vehicle, that damage would not be covered under warranty. This information was supplied in 1992. Subaru Customer Service and Public Relations did not answer several faxed requests to verify the information.

Suzuki, American Suzuki Motor Corporation
3251 E Imperial Hwy.
PO Box 1100
Brea, CA 92622-1100
714-996-7040

In 1992 the Suzuki Customer Relations Manager told us that the answers to our questions were "proprietary." This was clarified a bit by the additional statement that "American Suzuki Motor Corporation does not recommend any modifications or (non-Suzuki) parts or accessories." Two follow-up letters and two separate faxes for clarification went unanswered. Suzuki's Public Relations Department told us via telephone that "proprietary" was not a good choice of words. The intent of the original letter was to be clear that Suzuki will not support or endorse the installation of any after-market equipment and that any problems resulting from after market equipment will not be covered under warranty and should be resolved by the after market manufacturer.

Toyota Motor Sales USA, Inc.
Customer Assistance Center
19001 South Western Ave.
PO Box 1991
Torrance, CA 90509-2991
800-331-4331

Toyota says that the ECMs in all modern vehicles are easily damaged by electromagnetic radiation from high-power radio transmitters. The resultant problems could affect the operation of vital vehicle functions. The proper and safe operation of the vehicle could be compromised if any of the following situations, and possibly others exist: the transceiver is not type accepted; the power or antenna cables radiate RF current; the routing of the power or antenna cables results in inductive or capacitive coupling; transmitter, feed line and/or antenna inefficiencies results in an unacceptable level of RF radiation exposure to the ECMs; the SWR is unacceptably high or the antenna ground plane is inadequate.

Toyota has prepared an installation guideline entitled "Two-Way Radios In Toyota Vehicles." In addition to the potential

Society of Automotive Engineers
EMR and EMI Standards Committee

The automotive industry is quite concerned with automotive EMC issues. The Society of Automotive Engineers formed the EMR and EMI Standards Committee to coordinate and supplement the EMC work being done by individual manufacturers. This committee is still in existence today, with representatives from automobile manufacturers, component suppliers and consumer groups. Ed Hare W1RFI represents the ARRL and Amateur Radio officially on this committee, but there are several hams on the committee who also understand the radio communications aspects of the work and projects of the committee. The committee is publishing a new set of EMC standards for automobiles. SAE J55—which pertains to complete vehicle EMC tests, was first published in 1957. The document established a test method and limits for broadband radiation from ignition systems. It has been revised several times over the years to keep pace with changing technology.

The latest version, now in printing, includes a new test to measure the amount of RF energy and noise picked up on an antenna mounted on the vehicle. It also includes test methods for measuring the immunity of the vehicle to strong RF fields. Three immunity tests are included: off-vehicle radiation source, on-vehicle radiation source and a bulk-current injection test method. Of particular interest to amateurs is the on-board transmitter simulation test method. The amateur bands from 1.8 to 1300 MHz are included; power levels are typical of commercially available equipment for each band.

A second series of documents, SAE J1113, also in preparation, includes emissions and immunity tests for modules and compo-

nents used in vehicles. Originally, this document only contained immunity tests, but the scope was expanded during development to include emissions as well.

A task force of the SAE EMR Standards Committee is developing test methods that will have a significant impact on the design of integrated circuits (ICs) used in vehicle electronics. Improvements in the layout of ICs have been defined that will reduce the RF energy the microprocessors and other digital ICs will emit. While the focus of this standard is on automotive needs, it can be applied to ICs for all applications.

The work of the SAE committees is international in scope. SAE delegates represent the United States on two international standards committees that are chartered by the UN. The Special International Committee on Radio Interference (CISPR, from the French version of the committee name) deals with interference to radio reception. A working group of the International Standards Organization deals with the immunity characteristics of automotive vehicles. The SAE standards and requirements are closely aligned with the international standards.

It has taken several years to complete the work that went into these standards. The needs of Amateur Radio and other on-board radio services have been seriously considered as these standards have been developed. As manufacturers' engineering teams use these standards to develop new automobile designs, many of the problems reported by hams will no longer be a major concern.

—Poul Andersen, K8JOF, SAE EMR and
EMI Standards Committee Chairman

---

problems mentioned above, the guideline states that the maximum output power of the transmitter must be 100-W or less; all installations and operating instructions provided by Toyota and the transmitter manufacturer must be followed; the antenna must be installed as far away as possible from all ECMs or other on-board sensors: the antenna ca-

bling must be routed no closer than 20 cm (7 7/8 inches) from the ECM or sensors; all antenna and power cabling must not be routed alongside or near the vehicle's wiring harness, preferably crossing vehicle wiring at right angles; the antenna should be adjusted to obtain the lowest possible standing-wave ratio. All of this notwithstanding, they did emphasize that any damage caused by higher-power mobile radio is specifically excluded from warranty coverage.

Owner's manuals from 1993 and earlier published a 10-W power limitation. This is being deleted from 1994 manuals and the limitations in the guideline take precedent over those few 1994 manuals that have not been updated.

Customers or dealers having a problem can contact the Customer Assistance Center to obtain help or a copy of the installation guidelines.

Volkswagen of America, Inc.
Corporate Technical Services
3800 Hamlin Rd.
Auburn Hills, MI 48326
313-340-4723

All Volkswagen designs are thoroughly tested for EMC at their test facility in Wolfsburg, Germany. In addition, the vehicles are extensively tested in the field for RF exposures to fields greater than 120 V/meter. At the present time, the minimum requirements for passing their tests is 120 V/meter for 3 to 30 MHz and 80 V/meter for 30 to 1000 MHz. Critical safety components are tested over a wider range of frequencies at higher field strength. They have no reports of problems from using radio equipment in their vehicles. Their Corporate Technical Services Department is confident that they can advise hams that encounter problems.

Volvo Cars of North America
Consumer Affairs Department
Rockleigh, NJ 07647
201-784-4525

Volvo builds and tests vehicles to several international EMC standards. They feel that their cars are generally safe regarding immunity against EMI from properly installed transmitters. Volvo tests their vehicles with transmitters with frequency range from 1.8 MHz to 1 GHz, using 10 different antenna locations. Volvo requirements are that the vehicle performance shall not be affected by transmitters in the 200-W range and that no

system in the car shall be damaged by a field strength of up to 200 V/meter. This information was received in 1992. Several requests to verify the information for current model years were not answered.

LAB NOTES:
Mobile Installations and Electromagnetic Compatibility. Prepared by the ARRL Laboratory Staff (e-mail: tis@arrl.org)

BY ED HARE, W1RFI, ARRL LABORATORY SUPERVISOR

Q I read your QST article on mobile electromagnetic compatibility (EMC), but it left a lot of questions unanswered. It didn't tell me how to install a radio in my car, or even which cars are compatible with Amateur Radio equipment. Can you help?

A The article was not intended to be a technical discussion. We summarized the manufacturers' policies so that hams could make a decision about which vehicles were compatible with their Amateur Radio interests. As you might have noticed in the article, I was frustrated by some of the incomplete or evasive answers, too.

The manufacturer of a vehicle is still the best expert on how that vehicle performs, including its electromagnetic compatibility. If possible, select a manufacturer and dealer who offers technical support if you run into difficulty. In this regard, there were some clear winners (and losers) in the article.

Q Where do I start?

A Start with your dealer. Ask about service bulletins. Unfortunately, this process doesn't always work. Several members called me to get the correct Technical Service Bulletin numbers that the dealers and factory customer-service staff said didn't exist!

Also, ask your dealer about fleet vehicles. Some manufacturers have special modifications for vehicles intended for sale to police, taxi companies and other users who will be installing radios in the cars.

Q I was quite impressed with the auto manufacturers who had booths at the Dayton HamVention. Does this mean that I can't go wrong if I buy one of their cars?

A Oh, if only it were that simple. Unfortunately, each manufacturer has a number of different models, all of which have different options. The possible combinations can result in unexpected problems. If problems do occur, it is often not possible to engineer changes after the design is complete. What I'm saying is that even the good guys can make mistakes and the help they can offer is sometimes limited.

Q How can I tell which car is compatible? Does the ARRL maintain a database?

A No, we don't have a database. There are hundreds of different models in each model year and we have reports on only a handful. Most hams don't write the ARRL to report problems, and almost none write to tell us they are not having an interference problem.

Q But I still need to buy a new car. Is there anything I can do?

A Yes—you can test the cars before you buy them! EMC problems with vehicles come in two flavors—interference from the vehicle and interference to the vehicle. The easiest thing to check is interference from the vehicle. When you go to a dealership, bring a battery operated receiver that covers the frequency ranges you want to use. You can use a built-in whip antenna, but a mag-mount antenna will yield more reliable results.

Start the vehicle and turn on the receiver. Tune across the entire frequency range. You are looking for two types of signals: broadband noise and discrete spurious signals. Turn on all the accessories (wipers, turn signals, air condi-

tioning, and so on) one at a time or in combination. Have the dealer drive the car for a mile or so, just to see if there are any problems apparent while the vehicle is in normal use.

Q I hear a little noise and I found a few spurious signals. Does this mean I should not buy the car?

A Maybe not. Amateur Radio applications are only part of your decision. You may be able to live with some noise to get an otherwise fine car. Compare the amount of noise against your intended use of the radio. If you are going to use it to talk through a powerful, local repeater, the repeater will probably be strong enough to mask the noise. If the spurious signals fall on frequencies not used in your area, you may not care.

Q Okay, I can live with the receive problems. What about transmitting?

A This is a bit harder. The only way to conduct a test is to actually transmit. You will need to talk this over with the dealer. The dealer will not allow you to do a permanent installation, but you should be able to use a mag-mount antenna and external battery (I use a deep-discharge marine battery) for your tests. You may run into other snags when you hardwire the installation, but they can usually be fixed.

Q The dealer said "no way" when I asked if I could transmit. Is this common?

A Well, it isn't rare. This answer should tell you something, though. If they're not willing to work with you when you're in their showroom with a pocketful of money, they may be even less congenial if you have EMC problems later on. If you can obtain a copy of the car manufacturer's transceiver installation guidelines (if there are any), they may help you convince the dealer.

Q The dealer finally allowed me to try a transmit test. When I keyed the rig on 2 meters, the windshield wipers came on! Is this serious?

A It is a good thing you didn't try it using high-speed Morse code! Actually, compatibility problems come in a wide range of severity, from nuisances (like what you discovered) to more serious glitches. The manufacturers usually take care of the serious problems, but if minor malfunctions are found late in the design cycle, they may not be corrected.

Q Well, is it time to try another car?

A I would first experiment with the placement of the mag-mount antenna. You may find that it works well in a different place, especially toward the rear of the vehicle.

Q Bingo! During my test I had the antenna on the hood When I moved it to the center of the roof, the windshield wipers stopped misbehaving. Does this mean I can buy this ear?

A It will probably be okay. Just make sure you follow the car manufacturer's guidelines when you install the radio.

Q Some of the things in the installation guidelines are a real pain. Why can't I just follow "good engineering practice" in my installation?

A Manufacturers have worked out the proper installation procedures in great detail, even citing the locations close to the electronic control module (ECM) and wiring that should be avoided. Perhaps even more important, especially for vehicles that are still under warranty, you must follow the manufacturer's guidelines if you expect to obtain any support from a manufacturer or dealer. Even if what you do is technically correct, the manufacturer may choose not to support an installation that was not done their way.

Q Should I avoid cars that don't have published installation guidelines?

A Not necessarily. Even though some of the customer-service people I contacted didn't know it, all cars are subjected to EMC testing. After all, the manufacturers don't want their cars to fail when they're driven past a high-power transmitter site.

Transceivers can be installed in most cars. Even so, you may want to avoid manufacturers who state that it's not acceptable to put radios in their cars.

Q When I'm installing the radio, where should I connect the dc power leads?

A The positive and negative leads from the transceiver should connect directly to the battery with fuses in both leads. Route them away from any other wiring in the vehicle. This usually means drilling a small hole in the firewall between the engine and the passenger compartment and running the wires through a rubber grommet (to protect their insulation from sharp edges). If the wiring or antenna lead must pass near vehicle wiring, route it at right angles to the wiring. It is best to run the power and antenna leads as close to the vehicle chassis as possible.

Q Why can't I find power inside the car?

A It's often difficult, especially in modern cars, to know which power sources in the passenger compartment can safely carry the high-current loads of a transceiver. If you choose a source that also powers one of the microprocessors, the transmit current could cause problems. In addition, some transmitters are not well bypassed and some RF energy can appear on the power leads. This is bad news for microprocessors. Even grounding isn't as simple as it seems. With all of the plastics used in modern vehicles, not everything that looks like ground is really ground. It could be "floating" or even be part of the wiring for one of the vehicle's sensors. Even if you do tie into the chassis, it is possible that the section you've chosen forms part of the return path for one of the control modules. Generally, you're better off running the wires to the battery, especially for high-power installations.

Q I understand. I've been curious, though; why do I need two fuses at the battery? The radio has a fuse. Isn't that enough to protect the radio?

A The fuses at the battery are not just to protect the radio; they also protect the car. If a short circuit were to develop in the wiring, the radio fuse would not blow and the wiring could start a fire. Fusing the radio's hot lead at the battery could prevent this. I've received a lot of questions about the fuse in the negative lead. In most cars, there is a heavy conductor between the negative battery post and the engine block. This conductor is used to carry (among other things) the high current drawn by the engine's electric starter motor. If this negative lead were to fail (corrosion around a battery can take its toll), the starter current would flow through the transmitter's negative power lead! This wire is not rated to carry the high current of the starter. Fusing the radio's negative lead will prevent the wire from overheating and protect the transmitter installation. (I sure wish I had put this explanation in the original article; it would have saved me hours on the telephone!)

Q Where should I put the antenna?

A First, use only high-quality coaxial cable for the antenna feed line. This feed line should be as short as possible and located away from other wiring in the car. Most manufacturers' guidelines also say that the antenna should be operated with an SWR of 1.5:1 or less. This is not usually critical, although if the feed line is not well shielded, the leakage from the line will increase with SWR. (The most important reason to honor all conditions of the installation guidelines is that you may need to do so to obtain the support of the manufacturer.)

Most of the installation guidelines specify that the antenna be located on the roof, or at the rear of the car. Be prepared to experiment with the location. The mag-mount antenna you obtained earlier will help you do this, even if you plan to use a different mount for a permanent antenna.

The transceiver should be located away from vehicle electronics and wiring. Make sure that it will not interfere with the operation of the vehicle. One space often overlooked is the deployment area of the air bags! Don't put anything where they will interfere with the car's safety features. If you're not sure where after market equipment can be safely installed, ask the dealer.

Q My VHF installation works perfectly. However, I followed similar guidelines for HF and found that I had some pretty strong noise on receive. Some of this noise goes away when I connect the radio to the test battery I used earlier. Any ideas?

A It sounds like some of the noise is coming down the power leads. To cure this, first try ferrite chokes on the positive and negative leads. Obtain two FT-140-43 ferrite cores [2] and wrap about 10 turns of each wire onto the core. (Fewer turns may not work, and if you substitute a different ferrite, make sure it's suitable for HF—unknown materials usually don't work!) You can also try installing 0.01-μF capacitors from the positive lead to the negative lead, or from the positive lead to chassis ground (or both). The capacitors may work best with or without one or both of the ferrites.

Q That did improve things, but I still hear noise when I connect the antenna.

A Some of this may be coming from outside the car. First, see if the problem goes away when the car is turned off. If it does, it's being radiated by something in the car. Once it is radiated, it can't be filtered at the receiver, so you'll have to locate the source and correct it there.

There are a number of possible noise sources in a car. Listening to the nature of the sound may give you a valuable clue. It may sound similar to the digital noise you often receive near a home computer system or video game. If so, one of the vehicle's ECMs may be radiating. It may have the characteristic whine of electric motor noise. Determining which motor is operating may be fairly easy. For example, an electric fuel pump can be noisy. If the whine occurs all the time and varies with engine rpm, it's probably the alternator. It's also possible that an electric motor is defective, especially if the vehicle passed the noise tests you gave the demonstrator model in the lot. If so, the dealer can change the motor.

If you hear a fast pop-pop or buzzy sound that increases to a whine with the car's engine, you may have ignition noise problems. These are usually more noticeable on HF, although severe cases can bother VHF FM. If your receiver has a noise blanker, this may be effective at reducing or eliminating ignition noise.

If you didn't hear ignition noise when you checked the car at the dealership, it may have developed an ignition problem. First, check to see that both the wires and the plugs are "resistor" types. (In some cars, using both at the same time may reduce engine power slightly; check with your dealer.) You may also have a bad plug or ignition wire. This is fairly easy to check by substitution. Problems such as worn rotors, a cracked ignition cap, and so on can create ignition noise. In severe cases, it may be possible to add bypassing to the distributor components or try shielded wires. In modern high-voltage, computer-controlled ignition systems, these types of cures can cause more problems than they create and should be done only by qualified service personnel.

Q   How much RF power can I run in my car without endangering the vehicle electronics?

A   There is no clear-cut answer to this one. You should be able to do a successful installation with 100 W transmitters on HF, 25 W on VHF and 10 W on UHF and above. At VHF and UHF, you may be able to get away with power levels as high as 50 W. Most of the time, these power levels do not cause permanent damage to the vehicle. I recommend that you think long and hard before wiring in a high-powered mobile amplifier; it may do permanent damage to sensitive vehicle electronics.

Q   Thanks for your advice. Should I run outside and "fix" my car now?

A   If the car is under warranty, leave the repairs to the dealer. If you have an older car and usually do your own repair work (you should see some of my beaters!), obtain a copy of the ARRL book, *Radio Frequency Interference: How to Find It and Fix It*. [3] The chapters on automotive, electrical and computer interference may all prove useful.

We welcome your suggestions for topics to be discussed in Lab Notes, but we are unable to answer individual questions. Please send your comments or suggestions to: Lab Notes, ARRL, 225 Main St, Newington, CT 06111.

NOTES

1. E. Hare, W1RFI, "Automotive Interference Problems: What the Manufacturers Say," September 1994 QST, page 51.

2. These are available from distributors such as Amidon, FairRite and Palomar. Check QST advertisers or the parts distributors listed in the 1995 ARRL Handbook.

3. Available from your favorite Amateur Radio dealer, or directly from the ARRL. See the ARRL Publications Catalog.

FDA Documents on EMI

CDRH MEDICAL DEVICE ELECTROMAGNETIC COMPATIBILITY PROGRAM

The Center for Devices and Radiological Health (CDRH) has regulatory authority over several thousand different kinds of medical devices, with thousands of manufacturers and variations of devices. Because of its concern for the public health and safety, the CDRH, part of FDA, has been in the vanguard of examining medical device EMI (electromagnetic interference) and providing solutions. Extensive laboratory testing by CDRH, and others, has revealed that many devices can be susceptible to problems caused by EMI. Indeed, the CDRH has been investigating incidents of device EMI, and working on solutions (e.g., the 1979 draft EMC standard for medical devices), since the late 1960s, when there was concern for EMI with cardiac pacemakers.

TYPICAL ELECTROMAGNETIC ENVIRONMENT FOR MEDICAL DEVICES

Electromagnetic compatibility, or EMC means that the device is compatible with (i.e., no inter-ference is caused by) its electromagnetic (EM) environment, and it does not emit levels of EM energy that cause EMI in other devices in the vicinity. The wide variation of medical devices and use environments makes them vulnerable to different forms of EM energy which can cause EMI: conducted, radiated, and electrostatic discharge (ESD). Further, EMI problems with medical devices can be very complex, not only from the technical standpoint but also from the view of public health issues and solutions.

CDRH MEDICAL DEVICE ELECTROMAGNETIC COMPATIBILITY PROGRAM

MEDICAL DEVICES AND EMI: THE FDA PERSPECTIVE

NOTE:
The following article is text only. The figures have not been included.

Don Witters

Center for Devices and Radiological Health
Food and Drug Administration
Rockville, MD 20850

THE EMI PROBLEM

An electric powered wheelchair suddenly veers off course; an apnea monitor fails to alarm; a ventilator suddenly changes its breath rate.

These are just a few examples of the problems that might occur when radiated electromagnetic (EM) energy interacts with the sensitive electronics incorporated into many medical devices. Over the years, many incidents of suspected electromagnetic interference (EMI) with medical devices have been documented. In addition, recent congressional hearings and media attention have heightened concern for the safe and effective use of devices in the presence of EMI. For medical devices the environment has become crowded with potential sources of EMI.

Because of its concern for the public health and safety, the Center for Devices and Radiological Health (CDRH), part of the Food and Drug Administration (FDA), has been in the vanguard of examining medical device EMI and providing solutions. Extensive laboratory testing by CDRH, and others, has revealed that many devices can be susceptible to problems caused by EMI. Indeed, the CDRH has been investigating incidents of device EMI, and working on solutions (e.g., the 1979 draft EMC standard for medical devices), since the late 1960s, when there was concern for EMI with cardiac pacemakers.

The key to addressing EMI is the recognition that it involves not only the device itself but also the environment in which it is used, and anything that may come into that environment. More than anything else, the concern with EMI must be viewed as a systems problem requiring a systems approach. In this case the solution requires the involvement of the device industry, the EM source industry (e.g., power industry, telecommunications industry), and the clinical user and patient. The public must also play a part in the overall approach to recognizing and dealing with EMI.

The focus of this article is to briefly outline the concerns of the Center for Devices and Radiological Health, FDA, for EMI in all medical devices with electrical or electronic systems, and the strategy developed to minimize these problems.

THE COMPLEXITY OF DEVICE EMI

As our society seeks new technology, medical devices can usually be found in the forefront. There is an ever-increasing use of electronics and microprocessors in devices of all kinds, across the vast range of devices: from relatively simple devices like electrical nerve stimulators to the more recent advances in imaging such as magnetic resonance imaging (MRI). In the medical industry there is a tendency toward more automation in devices to monitor patients and help perform diagnosis. Microminiaturization has revolutionized the medical device industry: smaller devices requiring less power that can perform more functions. At the same time, there is a proliferation of new communications technology: the personal communications systems (PCS), cellular telephones, wireless computer links, to name a few. With these advances are coming some unforeseen problems: the interactions between the products emitting the EM energy and sensitive medical devices. Even the devices themselves can emit EM energy which can react with other devices or products.

Electromagnetic compatibility, or EMC, is essentially the opposite of EMI. EMC means that the device is compatible with (i.e., no interference caused by) its EM environment, and it does not emit levels of EM energy that cause EMI in other devices in the vicinity. The wide variation of medical devices and use environments makes them vulnerable to different forms of EM energy which can cause EMI: conducted, radiated, and electrostatic discharge (ESD). Further, EMI problems with medical devices can be very complex, not only from the technical standpoint but also from the view of public health issues and solutions.

A brief overview of radio frequency interference (RFI) can help to illustrate some of the variables that make device EMI so complex and difficult to address effectively. In general, the strength of the EM field at any given distance from the source of the radiated signal (transmitter) is directly proportional to the radiated power of the transmitter and inversely proportional to the distance. The role of distance from the EM energy source is highlighted by Figure 2. The relatively low power cellular telephone creates a 3 V/m field strength at 1 m, while a more powerful handheld CB transceiver creates the same field strength at 5 m. Further, the high power TV transmitter creates this same field strength at a distance of 1000 m. It is easy to see then, at small distances from the radiator where EM field strength can be very high, even the best protected devices (i.e., with a high level of immunity) may be susceptible to EMI. However, the device may be susceptible to only some of the variations (e.g., frequency or modulation) in the EM energy. This is why some devices may be affected by a nearby transmitter of a certain frequency, and other devices at the same location may not be affected. Add to RFI the other forms of EMI and it quickly becomes apparent that devices can face a fairly hostile environment, which can ultimately affect the patient or device user.

FDA CONCERN WITH EMI

The consequence of EMI with medical devices may be only a transient "blip" on a monitor, or it could be as serious as preventing an alarm from sounding or inappropriate device movement leading to patient injury or death. With the increasing use of sensitive electronics in devices, and the proliferation of sources of EM energy, there is heightened concern about EMI in many devices. While the numbers of reports with possible links to EMI have been steady, these numbers are generally not indicative of the actual occurrence of incidents. Indeed, in investigating possible EMI-related problems it is usually the case that the EM energy which caused the event has dissipated (e.g., the EM energy source was shut off or removed from the area). Only through careful measurement and testing can the true nature of EMI sus-

ceptibility be determined. The complexity of the testing, and the vast range of devices encountered, make it a very difficult task indeed to address EMI.

The CDRH has regulatory authority over several thousand different kinds of medical devices, with thousands of manufacturers and variations of devices. The very nature of this range of devices does not lend itself to "generic" approaches. For example, an apnea monitor is very different from a powered wheelchair, in form, function, and configuration.

The EM environment that envelops the devices can vary widely, from the rural setting to the commercial setting, to the urban setting, and of course, the hospital setting. The International Electrotechnical Commission (IEC) has classified the EM environment into eight areas and defined the typical EM environment in each area. Within each area there are conditions for the location and power of local EM energy sources (e.g., transmitters), which if exceeded would result in higher EM field strengths. Figure 3 indicates the general classifications and the upper range of radiated EM field strength specified for each environment.

FORMATION OF THE CDRH EMC WORKING GROUP

Concern in the CDRH has led to the formation of an EMC Working Group. This group was charged by the Deputy Center Director, Dr. Elizabeth Jacobson, to:

> assess all device areas to identify EMC concerns;

> coordinate the development of a strategy to assure EMC in all appropriate devices;

> provide a focal point for actions; and

> keep the Center Director and his staff informed of activities involving EMI/EMC.

This initiative involves virtually all of the CDRH offices and functions. The formation and subsequent accomplishments of the Group have already had an impact on the regulatory approach, research, and interactions with the device industry.

The EMC Working Group has developed a draft strategy to address EMC concerns across all appropriate device areas. This involves awareness (and education), regulation, research, cooperation with other agencies and organizations, and coordination and cooperation with manufacturers and users.

CDRH has long recognized that the majority of devices likely do not have major problems with EMI. Nonetheless, there are some critical device areas where the threat from EMI could directly impact upon the life and well-being of the patient. Rather than implement additional burdensome requirements over a broad spectrum of devices, CDRH is focusing on those areas where EMI has an established presence, is problematic, or could affect the critical function of the device.

PLANS FOR DEVICE EMC

A comprehensive plan for addressing medical device EMC needs to focus on the primary aspects of device safety and effectiveness. Although many manufacturers in certain device areas have been addressing EMC for some time (e.g., cardiac pacemakers), based on discussions with users, manufacturers, and EMC test facilities there still appears to be a general lack of awareness of the EMI problem. Thus, one key element in our plan includes raising this awareness and educating the users, manufacturers, and regulators about EMC.

Awareness

The CDRH has always placed a high priority on providing information to the public. For example, when the CDRH developed information that some apnea monitors could fail to alarm due to EMI, an FDA Safety Alert [figure 4] was sent out to large numbers of clinicians and users of these devices, warning of

the problem and providing tips for the safe use of the devices. Following the extensive investigations into EMI with powered wheelchairs and motorized scooters, the FDA published an article in its Medical Bulletin, which goes to over 1 million clinicians, providing information about device EMI. In addition, a question-and-answer document was developed for the users of powered wheelchairs and motorized scooters.

Pre-Market

The pre-market approach to device regulation was charged to the former Bureau of Medical Devices by the 1976 Amendments to the Food, Drug, and Cosmetics Act. In the early 1980s, this bureau was merged with the Bureau of Radiological Health to form the Center for Devices and Radiological Health. Under the 1976 Amendments, and the more recent Safe Medical Device Act of 1990, CDRH has authority to require device manufacturers to submit information about the safety and effectiveness of their devices. EMI has implications in both the safe and effective use of devices. Thus, a central part of the strategy for dealing with EMC concerns is to address these concerns in pre-market submissions. In some device areas, notably the respiratory and anesthesia area, concern with EMI has evolved over a period of years because of problems with such devices as the apnea monitor. Indeed, there is a draft FDA standard for apnea monitors with EMC requirements that grew out of our investigations of EMI problems. This draft standard is presently undergoing public comment.

Because of the vast range of devices, and the time and resources it takes to develop mandatory standards, a more general approach is being planned to address EMC in all appropriate device areas with respect to the pre-market concerns. This approach includes the development of priorities and guidelines for pre- and post-market and research activities.

Development of the guidelines for the regulators and manufacturers have been proposed in phases, including:

a general guideline to address EMC across a broad range of devices which would be harmonized with prevailing national and international standards; and ultimately, specific guidelines tailored to concerns in each device area and developed in accordance with pre-market priorities for EMC.

Post-Market

For devices already in use, the post-market domain, plans are being formulated to address EMC utilizing the Good Manufacturing Practice (GMP) requirements [Title 21 Code of Federal Regulations (CFR) 820] and inspection guidance [FDA, CDRH Compliance Policy Guidance Manual 7382.830, 5/94]. There are also plans to gather information from the manufacturers of radiation emitting products, such as electronic article surveillance systems, to examine the implications for device EMI.

In addition, the collection of incident reports, mandatory in the cases of patient death or injury, is another major tool to assess the post-market use of devices. With the large numbers of devices being used today, and the steady number of incident reports, plans are underway to better distinguish EMI incidents from other types of device incidents. The plans involve building a separate database of carefully scrutinized incident reports, which would form the foundation that would grow with later reports. A system to separate and analyze EMI reports will serve as a resource in making decisions and setting priorities.

RESEARCH AND STANDARDS

Research and work with voluntary standards organizations have been ongoing in CDRH for several years. Present investigations include examinations of suggested EMI to cardiac pacemakers from digital cellular telephones, EMI to ventilator devices, and follow-up on powered wheelchair EMC. The CDRH laboratory is equipped to perform

these kinds of investigations and has the experienced staff to develop test protocols. Indeed, the CDRH work with powered wheelchair EMC has contributed greatly to draft test requirements and procedures for a national (ANSI/RESNA) and an international (ISO) standard.

National and international standards activities play an important role in medical device EMC, which is why CDRH has promoted and supported the development of voluntary EMC product family standards for medical devices and EMC requirements for device-specific standards. In addition to ANSI/RESNA and ISO, CDRH has worked with AAMI, the ANSI-Accredited Standards Committee C63, and the International Electrotechnical Commission (IEC). In many cases, the Center's EMC laboratory findings and environmental measurements are utilized in proposals and recommendations to these voluntary standards organizations. The Center has been particularly interested and active in the development of IEC 601-1-2, and has attempted to harmonize our recommendations with this document to the extent possible, given the FDA mandate to assure safety and effectiveness. The European equivalent of this standard will become especially important as of January 1996, when the European Community EMC Directive becomes effective. IEC 601-1-2 is an important step towards assuring EMC of medical devices; however, CDRH has some critical concerns about this document, and is participating in the development of the first amendment to this document.

WORK WITH OTHER AGENCIES

There are additional plans to work with other Federal agencies and professional organizations to promote medical device EMC. Present activities include participation in the EMC Risk Assessment project ongoing at the Walter Reed Army Medical Center. Engineers at Walter Reed have begun an ambitious program to document the incidents of EMI in devices and address solutions. CDRH scientists have brought laboratory data and a rich history of experience to the meetings with Walter Reed staff. In addition, CDRH is continuing its dialog with the Federal Communications Commission (FCC) to promote medical device EMC.

SOME ACCOMPLISHMENTS TO DATE

The CDRH EMC Working Group, and previous work on device EMC, have accomplished much in a short time frame. Chief among the accomplishments is the formulation of strategies to address EMC in all appropriate device areas. By taking a more comprehensive approach, the CDRH has been proactive in raising awareness and concern for EMC/EMI in devices. The EMC Working Group cooperated with AAMI to present a one and one-half day forum on medical device EMC. The objective of the forum was simple: make known the concern for device EMC, and provide a forum for interaction by the users, clinicians, manufacturers, EM source industries, the public, and CDRH to address the concern.

The EMC Working Group has also been busy assessing the various device areas in the pre-market domain to help in devising priorities for guidance development and laboratory testing. In addition, the Group has provided training for the CDRH staff about EMC, developed strategies, and made recommendations for CDRH/FDA policy toward EMC. Various members of the EMC Working Group have been taking the lead in activities outside the CDRH to address EMC in medical devices.

The laboratory investigation of powered wheelchair EMI, and subsequent standards efforts, illustrates that device EMC can be achieved through cooperation among CDRH, manufacturers, and users. Below is a brief overview of this work.

Experience with Powered Wheelchair EMC

CDRH became aware of suspected EMI in powered wheelchairs and motorized scooters in mid-1992. By late 1993 CDRH laboratory investigations and testing had revealed serious EMI reactions by these devices over a wide range of radio frequencies (1 MHz to 1000 MHz). The evidence indicated that these devices could experience incidents of uncontrolled movement or electromechanical brake release in the presence of moderate radiated EM fields (as low as 3 to 10 V/m). This was sufficient to warrant notifying powered wheelchair users, through user organizations, of the potential for EMI, and to solicit information concerning actual incidents. Further testing revealed that the EMI seemed to affect the control system of the powered wheelchairs resulting in electromechanical brake release and unintended wheel movement.

In many cases, motorized scooters utilize the same type of control systems as the powered wheelchairs. Thus, there was concern that the scooter devices could also suffer from EMI. EMC tests were performed on samples of motorized scooters. The results revealed that these devices could also exhibit EMI problems. Experience from EMC testing of other devices led CDRH researchers to develop testing procedures which fully challenge the devices. These procedures became the basis for the 1993 CDRH proposals to the RESNA and the ISO for EMC tests and requirements in their respective standards. The proposals were made to harmonize as much as possible with the IEC 801-3 standard (recently renumbered to IEC 1000-4-3) for radiated immunity testing. However, in the process of performing the laboratory tests, CDRH created unique procedures which take into account the relatively slow response time of powered wheelchairs. Through careful scrutiny of submissions of EMC test data by the device manufacturers, and verification testing by CDRH, it became clear that the procedures devised by CDRH were more accurate in determining EMI problems than the existing standard procedures.

Additional testing procedures were developed to examine the device response as the wheels were kept at a constant speed, to simulate normal movement of the wheelchair. Figure 5 represents the results of testing on one device (before modifications were made by the manufacturer). In this case the wheels were fixed at a constant speed of 30 RPM during the exposure of the device. Note that there are several places where the motion of the wheels deviated from the 30 RPM baseline, indicating EMI to the wheelchair. These tests were performed at the EM field strength of 20 V/m. This level was chosen because the device manufacturers had stated they could build devices immune to this level, which is approximately the field strength from a hand-held transceiver at 0.6 m (2 ft). Many powered wheelchair users utilize radio transceivers and cellular telephones for communications, any of which could be placed within this distance of the device's control system.

Following careful EMC modifications to the powered wheelchair by the manufacturer, with the appropriate shielding and circuit modifications, the same powered wheelchair was retested and found to be immune (no EMI reactions) across the entire frequency range (Figure 6). This demonstrated that these devices could indeed be made immune to 20 V/m. With such findings in hand, CDRH notified powered wheelchair and scooter manufacturers in May 1994 that future submissions for these type devices should address EMC in labeling and testing. Additional work with the RESNA subcommittee for EMC refined the original CDRH EMC test proposal and reduced the number of test points, to make the procedure more affordable to perform, without compromising the test reliability. The experience with powered wheelchair EMI demonstrates the ability of CDRH to work with the device manufacturers to recognize and address an EMI problem. Many of these device manufacturers were helpful in sharing information, providing samples, bringing together interested parties, and working towards a solution of the problem. CDRH was able to develop a new and more accurate test procedure in a relatively short time frame, building upon its years of experience in the laboratory and EMC testing of devices.

SUMMARY

There is still much work to be done to reach the goal of assuring device EMC across the broad range of devices. The CDRH EMC Working Group has been charged by the Deputy Center Director to continue this effort, which will likely last some time into the future and impact all electrical and electronic medical devices. Given the nature of the EMI problem, and the quick pace of technology, plans for this program must be dynamic and flexible. The very nature of EMI is complex, with large uncertainties in nearly every aspect. The CDRH approach will reflect these constraints and rely in large measure on the cooperation of all of the parties.

REFERENCES

American National Standards Institute (ANSI)/ Rehabilitation Engineering Society of North America (RESNA) Proposal Addition to ANSI/ RESNA WC/14 Electromagnetic Compatibility Requirements for Powered Wheelchairs and Motorized Scooters, draft version 1.5 (November 1994).

Antila S., Where Shoplifting Bolsters Profits, *New York Times*, Money section, (December 12, 1994) page 13.

Baskim R., Haywire, segment on CBS television show *Eye-to-Eye with Connie Chung* (December 1, 1994).

Bassen H.I., D.M. Witters, P.S. Ruggera, and J. Casamento, CDRH Laboratory Evaluation of Medical Devices for Susceptibility to Radio-Frequency Interference, Designers Handbook: Medical Electronics, Canon Communications (February 1995) pages 44–49.

Bassen H.I., P. Ruggera, and J. Casamento, Changes in the Susceptibility of a Medical Device Resulting from Connection to a Full-Size Model of a Human, Proceedings of the 14th Annual International Conference of the

IEEE Engineering in Medicine and Biology Society (1992) pages 2832–2833.

Capuano M., P. Misale, and D. Davidson, Case Study: Patient Coupled Device Interaction Produces Arrhythmia-like Artifact on Electrocardiographs, *Biomedical Instrumentation & Technology* (November/December 1993) pages 475–483.

Casamento J., P. Ruggera, D. Witters, and H. Bassen, Applying Standardized Electromagnetic Compatibility Testing Methods for Evaluating Radio Frequency Interference of Ventilators (in draft)

CDRH's War on Electromagnetic Interference, An Interview with Donald Witters, CDRH, *Medical Design & Diagnostic Industry*, Vol. 17 No. 2 (February 1995) pages 34–40.

Clifford K.J., K.H. Joyner, D.B. Stroud, M. Wood, B. Ward, and C.H. Fernandez, Mobile Telephones Interfere with Medical Electrical Equipment, *Australian Physical & Engineering Sciences in Medicine*, Vol. 17 No. 1 (1994) pages 23–27.

Dear Powered Wheelchair/Scooter or Accessory/Component Manufacturer letter, Susan Alpert, M.D. Director Office of Device Evaluation, CDRH (May 26, 1994).

Draft IEC 1000-4-3 Electromagnetic Compatibility (EMC)—Part Testing and measurement techniques—Section 3: Radiated, Radio frequency, electromagnetic field immunity test (Revision of 801-3), SC65A/SC 77B (1994).

Draft International Organization of Standards (ISO) EMC Working Group Proposal Electromagnetic Compatibility Addition to ISO 7176-14, draft version 1.0 (April 1995).

Electromagnetic Interference with Medical Devices, *FDA Medical Bulletin*, Vol. 24 No. 2 (September 1994) pages 5–6.

FDA Safety Alert: Important Tips for Apnea Monitor Users (February 16, 1990).

International Electrotechnical Committee IEC/TC or SC: TC77, Electromagnetic Compatibility Between Electrical Equipment Including Networks, Classification of Electromagnetic Environments (1991).

Kimmel D.D. and D.D. Gerke, Protecting Medical Devices from Electromagnetic Interference, *Designer's Handbook: Medical Electronics*, 3rd Ed. (1994) pages 34–43.

Medical Electrical Equipment, Part 1: General Requirements for Safety, Collateral Standard: Electromagnetic Compatibility, International Electrotechnical Commission IEC 601-1-2 (1993).

Nave M.J., Consultant on Call: The Case of the Talking EEG Machine, Compliance Engineering (July/August 1994) pages 65–70. Electromagnetic Compatibility Standard for Medical Devices MDS-201-004, An FDA Medical Services Standards Publication, Rockville, MD, U.S. Dept. of Health, Education, and Welfare, October 1, 1979.

Radio-Frequency Interference, Readers Respond, *Paraplegia News, Paralyzed Veterans of America* (September 1993) page 6.

Radio Waves May Interfere with Control of Powered Wheelchairs and Motorized Scooters, sent with Dear Powered Wheelchair/Scooter or Accessory/Component Manufacturer letter and Dear Colleague letter, James Morrison, Director Office of Health and Industry Programs, CDRH (September 20, 1994).

Ruggera P.S. and E.R. O'Bryan, Studies of Apnea Monitor Radio Frequency Electromagnetic Interference, Proceedings of the 13th Annual International Conference of the IEEE Engineering in Medicine and Biology Society, Vol. 13, No. 4 (1991), pages 1641–1643.

Ruggera P.S., E.R. O'Bryan, and J.P. Casamento, Automated Radio Frequency Electromagnetic Interference Testing of Apnea Monitors Using an Open Area Test Site, Proceedings of the 14th Annual International Conference of the IEEE Engineering in Medicine and Biology Society (1992) pages 2839–2841.

Ruggera P.S. and R. Elder, Electromagnetic Radiation Interference with Cardiac Pacemakers, DHEW publication BRH DEP 71-5 (April 1971).

The Safe Medical Devices Act (SMDA) of 1990, Public Law 91-4243. Proposed Standard for Apnea Monitors 21 CFR Part 896, published in the Federal Register Vol. 60 No. 34 (February 21, 1995) pages 9762–9771.

Silberberg J.L., Performance Degradation of Electronic Medical Devices Due to Electromagnetic Interference, Compliance Engineering (Fall 1993) pages 25–39. Public Hearing, 103 U.S. Congress, Information, Justice, Transportation, and Agriculture Subcommittee, Rep. G. Condit (CA) Chairman, "Do Cellular and other Wireless Devices Interfere with Sensitive Medical Equipment?," Rayburn House Office Building (October 5, 1994).

Tan K.S. and I. Hinberg, Radio frequency Susceptibility Tests on Medical Equipment, Proceedings of the 16th Annual International Conference of IEEE Engineering in Medicine and Biology Society (1994).

Witters D.M. and P.S. Ruggera, Electromagnetic Compatibility (EMC) of Powered Wheelchairs and Scooters, Proceedings of the RESNA '94 (Rehabilitation Society of North America) Annual Conference Tuning in to the 21st Century Through Assistive Technology (July 1994) pages 359–360.

ELECTROMAGNETIC INTERFERENCE (EMI) TESTING OF MEDICAL DEVICES

FDA/AAMI Forum on Medical Device Electromagnetic Compatibility (EMC)

Keywords: electromagnetic compatibility, electromagnetic interference

To promote awareness of the potential for problems stemming from electromagnetic interference (EMI) and cooperation toward solutions, CDRH and AAMI co-sponsored a 2-day forum entitled, "Electromagnetic Compatibility for Medical Devices." The forum was planned and initiated by the CDRH EMC Working Group, which was formed by the Deputy Director for Science for CDRH and charged with assessing medical devices EMI and coordinating CDRH solutions. The forum was conducted this past May in Anaheim, California, in conjunction with the annual AAMI conference. The goal of the forum was to present the concerns of CDRH about medical device EMC and strategies being developed to address these concerns. The concerns for device EMI have been heightened in recent years because of the increased use of sensitive electronics in many devices, and several reports of device malfunctions relating to EMI. This comes at a time when there is a dramatic increase in the number of portable radio transmitters, from cell phones to the "wireless" computer communications. OST representatives moderated the presentations by representatives from the medical device industries, the electromagnetic source (e.g., radio, cellular telephone, AC power) industries, clinical device users (clinicians and biomedical engineers), the legal profession, and the Federal Communications Commission. Feedback from the presenters and audience indicated that there was a spirit of openness and cooperation which many had not experienced in dealings with FDA. The proceedings will be published in FY 96.

EMC Testing of Implantable Cardiac Pacemakers for EMI from Digital Cellular Telephones

Keywords: pacemaker, cellular telephone, electromagnetic interference

Testing has been performed on 24 different implantable cardiac pacemakers for EMI from digital cellular telephones. This work was done in response to public health concerns based upon information that digital cellular phones could interfere with the normal operation of pacemakers. Laboratory results indicate that about eight of the devices tested reacted to a digital cell phone (i.e., U.S. TDMA digital, GSM digital, or MIRS digital). Most reactions were only with the MIRS cellular phone technology. All but one device ceased reacting to the cell phone when the phone was vertically beyond about 9 cm from the pacemaker submerged in saline. This distance is much less than the 15.24 cm (6 inches) recommended in the draft HIMA labeling for implanted cardiac pacemakers. However, one device continued to react even when the cell phone was at a distance of more than 36 cm. The test results also indicate that there are several devices in which EMI could not be induced. Further testing is needed to assess the validity of our findings on a wider range and number of pacemaker devices.

Figure 6 illustrates the exposure set-up for testing the pacemakers. A torso simulator (saline filled tank) was used for all tests. This set-up allows for the precise mapping of the area of interaction (if any) in a plane above the pacer generator surface. The pacer generator was placed at two distances (1 cm and 0.5 cm) below the saline surface to examine the effects of implantation depth on any EMI. In most cases, an actual cellular telephone was used to expose the pacer device. However, some exposures were done with a dipole antenna (a more standardized radiator source) to simulate the cellular phone.

The major factors that seem to correlate with the EMI of pacemakers include cellular phone pulse rate and output power, pacer device and manufacturer, pacer cardiac signal sensitivity, depth of pacer below saline solution surface and, to a lesser extent, the polarity of the leads (unipolar/bipolar) and phone orientation. The results also impact implantable cardiac defibrillators (ICDs) because of the many similarities between these devices. Indeed, very preliminary EMI testing indicates that ICDs are also susceptible to cellular phone EMI. The sample devices were obtained from voluntary loans by the pacemaker manufacturers. Each of the manufacturers who lent devices for testing also provided equipment to program and interrogate the pacers. They were also allowed to observe and comment on some of the preliminary laboratory testing. This work is being coordinated with a similar ongoing project at the University of Oklahoma.

Environmental Compatibility

Keywords: electromagnetic compatibility, electrostatic discharge, power quality, intended-use environment

FDA requires a reasonable assurance of safety and effectiveness for Class II and III medical devices in their intended-use environment. CDRH, historically, has addressed the issue of intended-use environment on a case-by-case basis. In recent years, CDRH has attempted to address the electromagnetic aspects of the environment in a more uniform and proactive manner. OST is now expanding this approach to encompass additional environmental factors.

OST began in FY 93 providing engineering support to postmarket evaluation of electrostatic discharge (ESD) incidents affecting medical device performance. Since then, OST has expanded its involvement in ESD issues with representation in the standards development process and with development of in-house testing capability.

In FY 95, OST initiated a program studying power line disturbances and their effects on medical device performance. As a first step, the current situation in a representative medical facility was assessed. OST conducted an electromagnetic interference and electromagnetic compatibility (EMI/EMC) risk assessment at Walter Reed Army Medical Center in cooperation with the engineering staff at the medical center. The purpose of this study was to determine the power quality environment within the medical facility. This documented the types of power quality problems that might be encountered by medical devices in a hospital environment. In parallel with the testing conducted at Walter Reed, OST staff conducted a survey of applicable standards and general literature on power line quality. The staff also investigated and evaluated test equipment that can be used in the laboratory to simulate such power line disturbances as voltage sags, swells, surges, and transients. OST established a liaison with the Electric Power Research Institute (EPRI). There are ongoing efforts to participate in the EPRI Health Care Initiative that addresses power quality in the health care industry.

Over the long term, medical device performance under actual conditions of use is expected to improve. Experience with EMC has shown that, as the industry is made aware of problems with interference and is given guidelines regarding needed design specifications, medical device performance has improved.

In FY 96, this project will develop a test protocol for evaluating the performance of selected medical devices subjected to typical power line disturbances.

Ventilator Testing for Susceptibility to Radiofrequency Interference

Keywords: electromagnetic interference, ventilator

During the last year, OST scientists have received reports of ventilator malfunctions due to radiofrequency interference (RFI) generated by various radio sources. The environment is increasingly filled with radiofrequency (RF) sources from personal transceivers, cellular telephones, mobile radios, and fixed-base station transmitters. In response to these reports, OST scientists embarked on a program to test ventilators for susceptibility to radiated RF fields. Tests were performed using methods described in the IEC 1000-4-3 standard for electromagnetic compatibility (EMC) testing and measurement techniques. A servo-ventilator was tested in a gigahertz transverse electromagnetic (GTEM) cell using standardized methods for interference of function from RF electric field strengths of 3 and 10 volts per meter. The test RF fields from 30 to 1000 megahertz (MHz) were modulated using three different waveforms: 0.5 Hz square wave, 1 kHz (kilohertz) 80% amplitude modulated (AM), and unmodulated constant wave (CW). The performance of the ventilator was monitored by a ventilator tester that measured breaths per minute, tidal volume, minute volume, and relative pressure. A personal computer controlled the strength of the RF electromagnetic fields (EMI) and collected data from the ventilator tester. Interference patterns were compared to determine which modulation caused the most ventilator performance degradation. The ventilator was most susceptible to RF electromagnetic fields when it was tested in the infant mode. For 10 V/m exposures using 1 kilohertz (kHz) 80% AM modulation, complete cessation of operation was observed for a narrow frequency band between 157 to 168 MHz. The data shown includes 20% error brackets. RF electric fields at frequencies that cause the ventilator performance to degrade to values outside of the 20% performance limits are considered failures due to RF EMI. Significant performance degradation also occurred from 130–345 MHz, 290–295 MHz, and 820–865 MHz.

In the infant mode, when exposed to 10 V/m RF electric fields, the ventilator tidal volume performance varied beyond the 20% limits for the 0.5 Hz square wave modulation in the 680–825 MHz frequency range. The 1 kHz, 80% AM modulation created the most interference with normal operation from 130–345 MHz. Complete ventilation cessation occurred during exposures to frequencies of 157 to 168 MHz. The unmodulated RF did not interfere with the ventilator as much as the other modulations that were used. Different modulations would sometimes cause the ventilator to perform differently at given frequencies. At 160 MHz, the 0.5 Hz modulation caused the relative pressure to increase, whereas the 1 kHz modulation suppressed the relative pressure. The unmodulated RF caused broader interference patterns below 350 MHz. Of the three sides tested, the front exposure yielded the greatest sensitivity.

The duration of exposure at each frequency did affect the test results. The ventilator tester required two breaths to acquire data at each frequency tested. The pilot testing was done at 5 MHz intervals. For the case of the ventilator in the infant mode, this corresponded to a dwell time of 11 seconds at each frequency tested. When the capability to measure relative pressure was added to the testing protocol, the time to acquire all of the data at each frequency increased to 16 seconds. When detailed testing was conducted at 1 MHz intervals, frequent alarm sounding occurred, and failure magnitudes were significantly greater than those measured during the pilot testing.

The GTEM cell and the computer-controlled EMC testing facilities in CDRH provide the capability to effectively evaluate the performance of medical devices, such as ventilators, in the presence of RF fields. Devices can be exposed to RF fields with different amplitude modulations and field strengths to determine how they react to the different parameters. RFI testing in the GTEM cell provided a simple, repeatable means of testing the ventilator over a broad frequency range. A reviewer guidance for environmental and electromagnetic compatibility testing for respiratory devices resulted from this work.

Testing of Hearing Aid Interference from Digital Cellular Telephones

Keywords: electromagnetic interference, hearing aids, cellular phones

As a result of recent reports of hearing aid interference from digital cellular telephones, the Office of Device Evaluation requested that OST conduct a laboratory study to quantify the extent of the interference. Seven "in-the-ear" (ITE) hearing aids were tested for audible interference using a stethoscope earpiece and tubing at various distances from two types of digital cellular telephones. Interference, which took the form of a buzzing tone, was perceived up to 47 cm away from the transmitting phone. Within a few centimeters of the phone, the induced buzzing tone precluded normal use of the phone. The level of induced interference in the ITE hearing aids as well as six behind-the-ear (BTE) hearing aids was quantified using a sound pressure level (SPL) meter with a calibrated microphone. The highest interference measured was slightly above the normal threshold for pain within a centimeter of the phone.

These preliminary results demonstrate that there is not an immediate health risk to hearing aid wearers from cellular telephone use in their general vicinity. However, results also show that the hearing aids tested could not be used in conjunction with GSM and US TDMA digital cellular telephones. In some cases, potentially dangerous tones were produced within a couple centimeters of the phone. Further testing will be conducted on a larger representative sampling of hearing aids in order to verify these results. Information learned from these experiments will be incorporated into revised testing protocols in future hearing aid standards and guidances.

NEW DOCUMENTS AVAILABLE TO HELP RESOLVE MEDICAL DEVICE EMC PROBLEMS

Two U.S. national medical device electromagnetic compatibility (EMC) documents were recently published which aid clinical/biomedical engineers to address the device EMC issue in an informed and consistent way: a technical information report published by AAMI, and a recommended practice for ad hoc EMC testing of devices. These documents were developed through consensus processes (similar to the ANSI standards process) with input from industry, government, users and independent parties. The documents are written in a straight-forward and easy to understand manner, making it easy for all those involved with the manufacturer and use of medical devices to understand the concerns about electromagnetic interference (EMI) and ways these concerns can be addressed.

The Association for the Advancement of Medical Instrumentation (AAMI) published Technical Information Report (TIR) 18, Guidance on electromagnetic compatibility of medical devices for clinical/biomedical engineers—Part 1: Radiated radio-frequency electromagnetic energy. TIR 18 provides information and guidance on medical device EMC to clinical engineers and other biomedical personnel to help them evaluate the radiated radio-frequency (RF) electromagnetic environment in their individual health-care facilities, and implement actions needed to minimize the potential risks associated with electromagnetic interference (EMI) problems. This document contains sections on assessing the RF environment in the clinic, developing policies, examples of device interactions, and a bibliography of references for device EMC.

TIR No. 18 -1997 is available from:

Association for the Advancement of Medical
  Instrumentation
3330 Washington Blvd, Suite 400
Arlington, VA 22201-4598
Phone: 1-703-525-4890 or 1-800-332-2264
FAX: 1-703-276-0793
http://www.aami.org

The American National Standards
Institute (ANSI) Accredited Committee C63
(EMC) published C63.18, Recommended
practice for an on-site, ad hoc test method
for estimating radiated electromagnetic immunity of medical devices to specific radio-
frequency transmitters. C63.18 is intended to
provide an inexpensive, relatively repro-
ducible test method for estimating the radiated RF electromagnetic immunity of medical
devices to available, portable RF transmitters

that might be operated in the vicinity. This
document provides a test method that can be
performed by clinical and biomedical engineers to improve reproducibility and inter-
comparison of test results. It also provides
information to facilitate development of policies and procedures for managing the use of
specific RF transmitters within specific areas
of a health-care facility.

ANSI C63.18 (IEEE product number
SH94556) is available from:

The Institute for Electrical and Electronics
  Engineering (IEEE)
In the United States and Canada:
  1-800-678-IEEE (4333)
Outside the United States: 1-732-981-0600
FAX: 1-732-981-9667
http://standards.ieee.org/catalog/electro-
  mag.html

Index